The Tigerfibel
...sooo'ne schnelle Sache

Tiger primer... what a fast thing!

Pzkpfw VI "TIGER" I (Sd.kfz.181)
The Original Tiger Tank Manual — An Authentic Translation

TRANSLATED BY WULF-D. BRAND

2nd Edition, 2014
ISBN: 978-0-938242-05-5
Printed in the United States of America

PORTRAYAL PRESS
1 Broad Street, Suite 944
Branchville, NJ 07826

http://www.portrayalpress.com

Copyright© 2014 BY WULF D. BRAND
All rights reserved. No part of this publication may be reproduced, distributed, or transmitted in any form or by any means, including photocopying, recording, or other electronic or mechanical methods, without the prior written permission of the publisher, except in the case of brief quotations embodied in critical reviews and certain other noncommercial uses permitted by copyright law. For permission requests, write to the publisher, addressed "Attention: Permissions Coordinator," at the address above.

PORTRAYAL PRESS

2

For Platoon-leaders and Tigermen

D 656/27

The Tiger

Man - what a machine!

Handles like a car!

At Your fingertips are

- 700 horsepower
- 60 tons of steel
- 30 mph top speed on tarmac
- 15 mph in rough terrain
- 13 ft under water

Will shoot everything to pieces!

In the course of one day at the northern front

Lieutenant M. killed   38 T 34

         In recognition

         he received the

         "Ritterkreuz"

This Tiger tank received the following hits within a period of six hours at the southern section:

227 hits by anti-tank rounds

14 hits by 52mm shells

11 hits by 76.2mm shells

None of the above penetrated its armor!

Within the tracks, rollers and links had been shot to pieces.
Two crank arms were no longer functional. Several hits of anti-tank gunnery laid directly on the tracks. The tank had rolled over three mines.
This tank negotiated another 60 km of rough terrain under its own power.

Will withstand anything!

Not to be victimized by outside force,

but singlehandedly abolished from within.....

D A N G E R lurks in the sump !

Remember then :

Read Your manual well, otherwise

Your Tiger goes to hell !

Motto: To study, the fool will take great pain
The Tigerman enjoys learning, not in vain

Moral: Even moralists and their preachings
are immoral at times, despite all teachings

The Tiger Tank Manual

Published on 8/1/1943 by

THE CHIEF OF STAFF OF THE TANK CORPS

CHIEF OF STAFF

TANK CORPS   Headquarters, 8/1/1943

I authorize the Tiger Tank Manual

Guderian

Table of Contents

Oscar the Offroader.................11
Fuel...............................12
Power..............................16
Water..............................17
Manual Inertia Starter.............22
On the move........................24
6 x Check Oil......................25
Oil Pressure.......................28
Waiting............................29
B.Engine...........................30
The four two bbl. carburetors......32
C.Drive Train......................35
D.Running Gear.....................40
Defensive Driving..................43
Engine Shutoff.....................48
Field Recovery.....................50
Loading for Transport..............52

Radioman Weenie Descrambled........53
The Radio Apparatus................55
Intercom...........................56

Barrelbum the Freeloader...........59
The Versatile Cannon...............61
Slow Response......................62
5 cures for jamming................64
A mule's barometer.................66
The machine gun barometer..........67
EN-RA-DRI-LI-Clear, magic recipee..68
Turret Trouble.....................69

Gunner Glasseye, always on target..71
The Notch..........................73
Eyeballing.........................74
Measuring..........................76
Seven Goodies......................77
Elvira gets shot...................79
Barrelbum always hits..............80
Barrelbum's Belly Button Rule......81
Sensible Use of Ordnance...........83
Knife or Fork?.....................85
The Lead...........................87
Centering..........................89

Tank Commander Speedy Smart........91
Order to shoot.....................92
Daily Meals........................95
The Cloverleaf.....................96
The Reticule Gap...................98
"Wanted"...........................99
The Anti-Goetz....................100
Tank Theft........................101

Cloverleaf Charts.................104

Armor Intelligence Eastern Front..115
Light Armored Vehicles............116
Medium Armored Vehicles...........128
Heavy Armored Vehicles............146
Older Designs.....................152

Armor Penetration Charts..........155

Explanatory Notes.................168
Translator's Note.................171

Driver

"OSCAR the OFFROADER"

You drive a tank which has few opponents worth mentioning, but also very few brothers. It is up to You wether the Tiger turns into a predator waiting to charge or into a heap of rubble.

Fuel

Motto: Oh friend, there are two sides to fuel, on the one hand it drives You, on the other You fly with it!

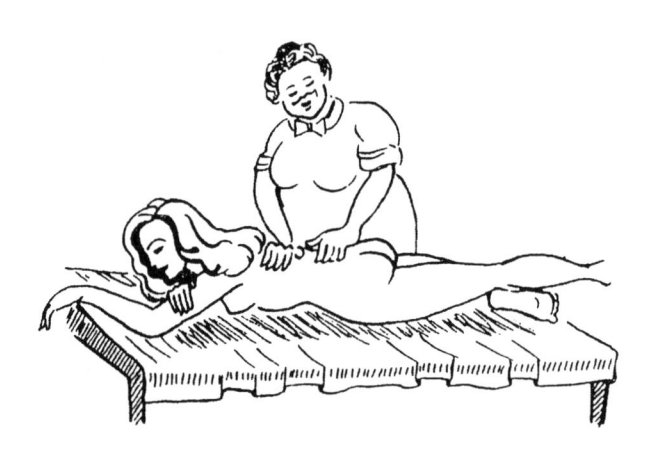

Fuel is a propellant:

If fuel evaporates and is then mixed with air and ignited in small amounts it will move all 60 tons of the Tiger along the road. It does so in many small explosions, much in the same manner as a child may roll a tire using a stick, touching the tire many times.

One Liter in the fuel reservoir will make the Tiger go a distance of 200 meters. The force of a giant is behind that motion, but it is distributed over half a minute just like a massage. And the Tiger likes that.

Fuel is an explosive:

If that same liter of fuel flows into the sump and not through the carburetor, then the fuel will evaporate by way of engine heat, mix with the air circulating in the engine bay and ignite at a hot spot or through a spark, all at once.

This one liter explodes Your Tiger in such a manner, that the engine and Your own roof fly up higher in the air than You can throw a rock. The giant's force is applied at once in a single knock out, which even Your Tiger cannot withstand.

T h e r e f o r e :

Motto: Minor details make a good job perfect,
A good job is no minor detail.

A runner will need two hours to warm up for the competition.
If he does not, the best equipment and the most vigorous training are worthless.

FOOD

GREASE

WARMUP

FUEL

6 x CHECK OIL

POWER

OIL PRESSURE

WATER

WAITING

STARTUP

The driver of a Tiger tank needs two hours to get his vehicle moving.
Otherwise it will break down because of a minor problem.
Prevention is easier than administering a cure. Therefore, before starting, pay attention to : FUEL - POWER - WATER - STARTUP - 6 x CHECK OIL - OIL PRESSURE - WAITING !

Therefore: Refuel - but do not spill,
Otherwise the Tiger will burn or burst.

Attention! If fuel is running low - immediately switch to reserve, if the fuel reserve is used up stop immediately and shut off the engine,
- 30 seconds worth of effort.

If You do not the fuel lines and the fuel pump run empty, after refueling no fuel will go through: Remove air filter and housing, remove access screws at the carburetor, prime with the electric fuel pump without causing a spill! Reinstall everything,
- One hour of work!

Close the lid for the fuel reservoir, but keep the vent open.

Otherwise the engine will have no pickup.

Keep fuel canisters and hoses clean, do not remove the strainer.

Otherwise the sightglass and jets will get dirty, both can be reached only with difficulty for repair.

Clean sightglass of dirt and moisture, do not damage the seals, if in doubt replace seals, insert properly, tighten cap nut.

Otherwise the Tiger will burn or burst.

Fuel line
and pumps must be checked for leaks. Fittings and lines must be thightened carefully. Do not cock the cone seal, it will rattle loose in the course of operation................................... Otherwise the Tiger will burn or burst.

Carburetor
The main jet must be cleaned with care. Blow air through the jet when finished. Check if the needle returns from the seat immediately. It must not hang up or be damaged........................... Otherwise, backfire at idle.

In case of fire
The warning light next to the driver's lookout will glow when the fire control apparatus is dispensing liquid. If it does not, the gunner must press the button on the fire extinguisher. You must immediately shut off fuel supply to the engine. Do not open to full throttle position as on other vehicles.................... Otherwise the gases intended to extinguish the flames will be removed by the blowers. The liquid dispensed by the fire extinguisher will not put out the flames unless it has been evaporated.

Extinguisher jets
must aim at the temperature sensors............. Otherwise, the apparatus will dispense all liquid at the first incident of fire. It is designed to extinguish for seven seconds each time, on five different occasions.

Temperature sensors
and lines must not be damaged when working on the engine. The fire extinguisher must be exchanged for a new one as soon as the pressure falls below 4 atmospheres. Check the fuses....... Otherwise You must use the manual fire extinguisher.

Cause of fire
is always fuel or oil in the sump. Check all the lines at once................................ Otherwise it will happen again!

|| *Moral:* If fuel is leaking from Your lines it will explode under Your behinds. ||

Power

Motto: He who treats his batteries with great care, right much he'll get in return - and that is fair!

They are Your best comrades,

They start Your engine as the bullets fly outside. They fire off Your shells, they siphon off the smoke! You can see in the dark, go straight ahead in fog, communicate amongst the loudest noise, talk and listen as far as 10 km away!

Give out an extra round often!

So that they can get really filled up!

Keep them warmhearted towards You!

A charged battery will freeze at -65°C, a discharged battery will already freeze at -10°C. That can happen very easily. Voltage and charge decrease rapidly as it gets really cold outside.

Give them some electricity,

if You have a surplus, so they can help out when You are running short.

Moral: There goes an old tale, who is filled up will not freeze pale!

Use distilled water or boil the water off. In the wintertime melted snow may be used in emergencies. Fill up until the plates are covered by one finger's width. Keep the clamps free of acid, tighten them firmly and cover them with grease.
Otherwise the battery will sludge up or go bad!

Use the manual inertia starter to turn and start the engine in the winter. That is a lot cheaper than taking out the batteries and recharging them! If voltage falls below 11 volts or the Tiger has been parked in the cold for a long time - take the batteries out and care for them! Otherwise they will explode.

12 volts! Do not let it fall under 11 volts. To check connect voltmeter between ground and fuse for high beam. Turn on high beams. When checking with the hydrometer, if mark at 1.285 = charged, 1.15 = empty.
Do not create a short circuit, do not change battery hookup.
Otherwise the plates will be distorted!

Motto: The power of water will work like a smooth breeze, that is, if You thought of antifreeze!

Water is a coolant,

Like a fresh shower, the water flushes through the case and removes the heat accrued by combustion and friction, taking it to the radiators. In addition, in the winter time it will preserve engine heat making it easier to start, much like a battery holds a charge.

The Tiger needs 120 Liters of water. He feels great at 85°C.

Water

If You are thirsty, give some water to the good animal too, and make it clean water. If You can get the coolant additive "Akorol", put it in, but be careful, Akorol is a poison. Like a hawk it will prevent rust and mineral deposits.

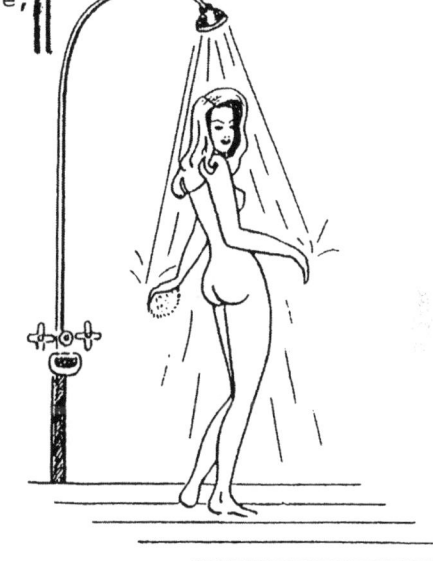

Water is an explosive,

When freezing to ice it expands by 10%. If the walls cannot give in, they are cracked under the enormous pressure. Ice is used to break iron and rock.

Those 120 Liters will then turn to 132 Liters, and they have no place to go. Not even in the Tiger's stomach!

T h e r e f o r e :
Fill up, until the water level touches the bottom of the strainer, check the hose clamps and lines, check especially on the bottom connecting the radiators.

95°C – attention! That is already too hot. The oil now ceases to lubricate and You can start looking for a new engine. Stop immediately and check:

1. Is there enough water in the radiators?
2. Are the blowers working?
3. Are the sliding clutches operating?
4. The oil cooler must be sealing tight!

Otherwise the engine will freeze up!

In the wintertime a grog is best mixed a little more on the stiff side. When it is cold, mix the water with antifreeze.

Otherwise the engine will burst!

Here now the infamous recipe from the ice bar at Dicken's Cold:

2 Liters of antifreeze per 3 Liters of water.
Down to -20°C that's: 48 Liters of antifreeze
                      +72 Liters of water
                      ───────────────────────
                      120 Liters

Down to -40°C that's: 72 Liters of antifreeze
                      +48 Liters of water
                      ───────────────────────
                      120 liters

Constantly check the cooling system, because antifreeze loosens all deposits. But do not mix antifreeze with the additive "Akorol".

Open the filler cap cover. Drain the water hot, through a rubber hose attached to the drain cock. Before refilling flush the cooling system and tighten the drain cock with gasket installed.
- After a week again drain hot, let sit for three hours, so that rust and sludge can settle, refill using a rag as filter; check and top off water each time after working on the engine.

Otherwise the radiator will leak.

‖ Moral: ‖ The Tiger guzzles the water down like You the schnaps when out on the town.

Motto: A cross country skier takes his time to prepare, if he starts in a haste he won't make it there.

Before the start the runner will check the seat of his shoes and ties with care. Even millimeters make a difference.

Before starting out, carefully check the position of all levers. One look must tell You if everything is okay. They are all essential for survival.

Fire extinguisher

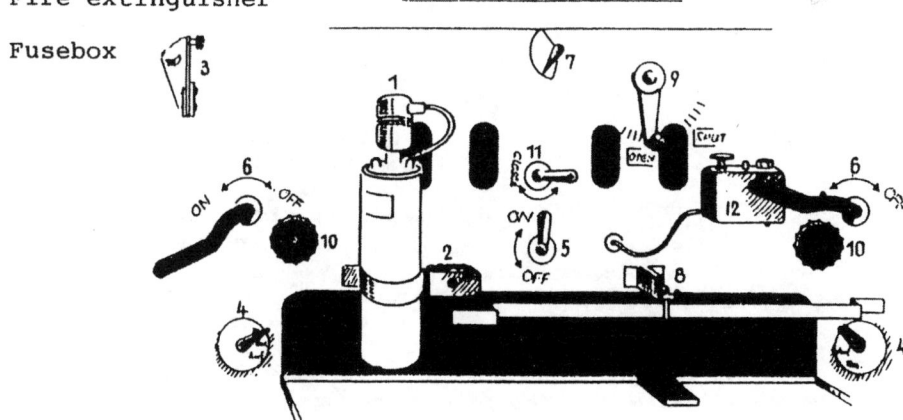

Fusebox

3.	Bottom plug	open	so that the gases can escape, at the same time keep rear cover open, so that fresh air can follow. Close only when driving in or through water, sludge or upon engagement. - **Otherwise the Tiger will burst!**
4.	Fuel valves	open	Fuel reservoirs must empty one after the other. If fuel runs out, immediately turn off the engine. - **Otherwise, see "Fuel".**
5.	Main battery switch	on	It turns off all appliances. - **Otherwise You cannot start!**
6.	Blower switch	"Land"	The blowers are turned off under water. - **Otherwise the engine boils!**
7.	Fuel vent	"Land"	While under water the tanks are vented into the engine bay. - **Otherwise to the outside.**
8.	Throttle #1	down	It regulates air flow in the duct between blower and transmission. - **Otherwise the transmission overheats.**
9.	Throttle #2	"Land"	It directs the hot air from the transmission to the blowers or into the engine bay. - **Otherwise the engine boils.**
10.	Throttle #3	open	It directs the hot engine air to the blowers. - **Otherwise the engine boils.**
11.	Vent flap	closed	Open only together with front hatch. - **Otherwise engine bay air enters the vehicle.**
	Fuel pump	on	So that the carburetor has fuel as You start. - **Otherwise the battery goes empty.**
	Directional lever	on "0"	forward = lever forward, reverse = lever back. - **Otherwise the tank moves upon starting.**

Ignition key	in	Do not turn, turn off other appliances. - **Otherwise the battery goes empty.**
Choke lever	forward	So that the mixture is enriched. Take the foot off the throttle when starting. - **Otherwise it will not start.**
Clutch	depress	So that the starter motor does not turn the transmission when cranking. - **Otherwise the battery goes empty.**
Starter button	push	Push longer, take a longer break between attempts at starting the engine. - **Otherwise the battery goes empty.**
Starter button	release	as soon as the engine starts, - **Otherwise the starter motor suffers.**
Choke lever	reverse	as soon as the engine runs smoothly and evenly. - **Otherwise the spark plugs foul.**
Throttle	touch	lightly for five minutes, so that warning lamp flickers. Do not race the engine. - **Otherwise it will hesitate.**
Clutch	engage	slowly, so that transmission and steering gear become warm to the touch. - **Otherwise no gear change.**
Throttle	push in	to warm up the engine. Increase engine speed to within 1000 and 1500 rpm. - **Otherwise the spark plugs foul.**

Manual Inertia Starter

In the wintertime...
the oil gets thick, so much so that it can hardly be called a liquid. Shafts bind in their bearings, pistons adhere to the cylinder walls. It takes a tremendous amount of force to separate these parts and move them until the engine oil gets warm and liquid. Although the Tiger can be started immediately at temperatures down to -20°C using the electric starter motor, use the manual inertia starter instead to crank over and start the engine. Save the batteries.

- Otherwise You cannot start when the bullets fly outside.

When it is very cold ...
one Tiger can warm up the other. The hot engine coolant in one engine is pumped into a cold engine. At the same time the cold engine is warmed up.
After this procedure, be sure to check that normal operation has been reinstated.

- Otherwise the Tiger will blow up.

Blowers...
must be shut off, so that the engine heats up faster.
Watch the thermometer carefully.

- Otherwise the engine will boil over like a pot full of soup.

Inject...
if the manual inertia starter does not start the engine.

- Otherwise You will lose faith and time.

In the tropics...
and in high summer conditions the blowers are set at high speed.

- Otherwise the engine boils.

Manual inertia starter...
accelerate with the hand crank in clockwise direction.
Swiftly engage the crank and hold it until the engine starts,
then immediately releasing the crank!
If the pinion does not mesh, repeat the engagement of the
crank. Do not engage the crank when the engine is running!

To exchange engine coolant...
A. Fill the hoses.
   1. Shut off the engine.
   2. Attach the hoses to red fittings.
   3. Shut off radiators through in line valves.
   4. Shut off blowers, remove radiator cap.
   5. Run the engine, hold up the hose,
      move the plunger at the open end until water comes out.
   6. Replace lost water and antifreeze.

B. Exchange the coolant.
   1. Shut off engine as soon as it reaches 60°C coolant temperature.
   2. Connect the hoses so that each one connects a red and a
      green fitting.
   3. Again close the valves in the coolant lines of the warm
      engine. Turn off the blowers, remove radiator caps.
   4. Run the engine at 2400 rpm, then at 2000 rpm. until
      the cold engine has reached a coolant temperature of 50°C.
   5. Shut off the engine, open the coolant lines,
      turn on the blowers, close radiator caps.

Selector...
on the blower transmission must be loosened and switched
back to position "increased cooling", then tightened.

|| *Moral:* : A little waltz, a warm drink for a while ||
|| even a frigid one will start to smile. ||

On the Move

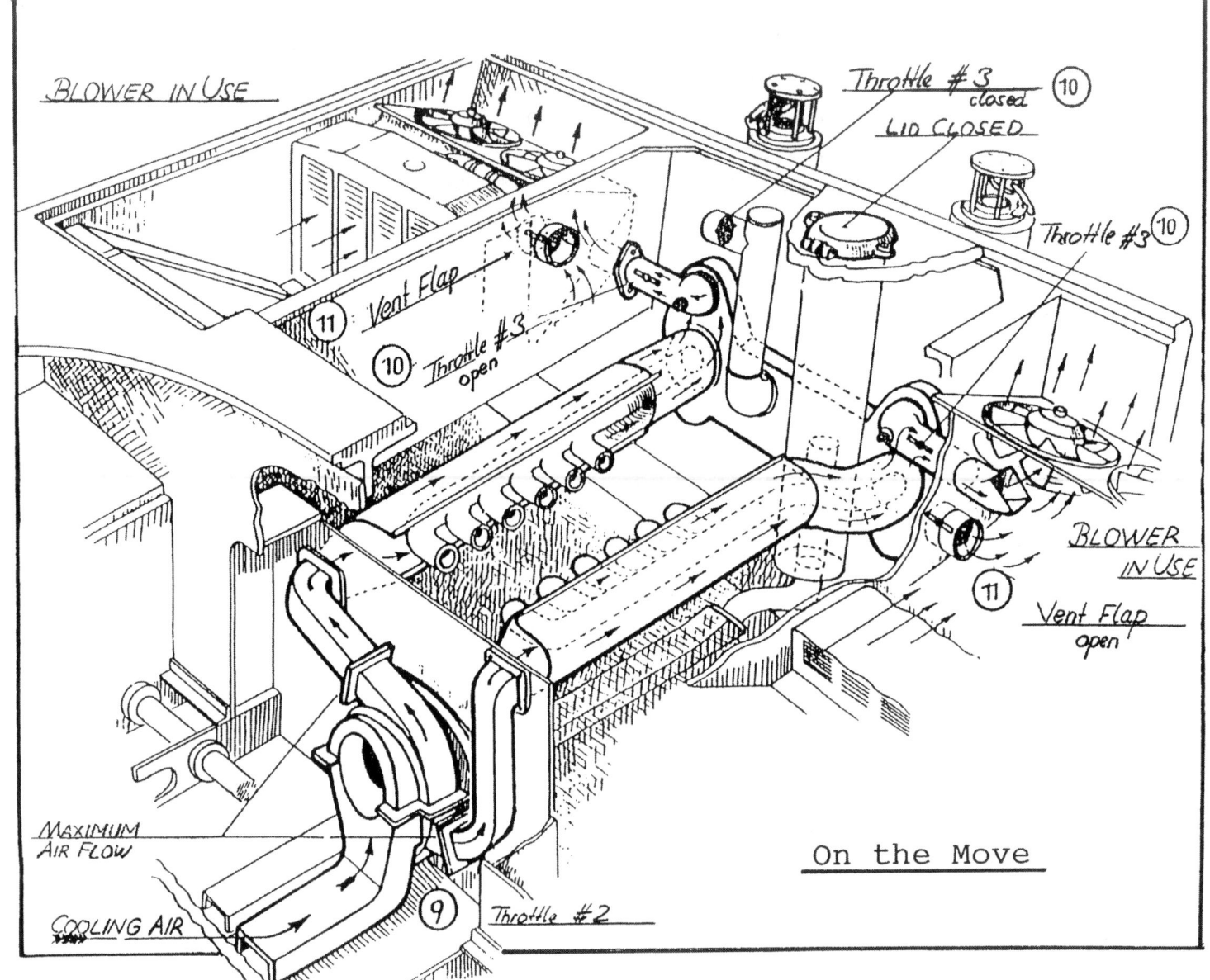

6 x Check Oil

Motto: In this case the oil prevents the sun's touch the Tiger likes oil very much.

Oil is a lubricant
Even just rubbing Your hands one against the other will cause them both to become hot. You need not even rub quickly or use much effort. But, if sufficient skin oil is placed between Your hands, they stay cool.

Your machine does 3000 rotations per minute with 700 horsepower behind it. It would get burning hot, all moving parts would freeze up, You could not move one kilometer, if oil did not take up the heat and flush it away. A low oil level is dangerous.

Oil is a combustible
If it leaks from Your lines, is thrown out by moving driveshafts, drips from damaged seals and mixes with fuel, it will burn furiously and set other puddles of fuel and the remaining sludge in the sump on fire.

Too much oil is dangerous.

Therefore:

Moral: It will freeze up if You don't lubricate, if You lose it, You will fumigate!

Fill where?	Fill what?	Fill how much?
1. Engine	28 L engine oil	Maximum level at upper mark, Minimum level at lower mark.

- **Otherwise** the spark plugs foul. You will use lots of oil and engines.

2. Transmission	30 L of transmission oil	Until the measuring rod just touches the oil level.

- **Otherwise You can neither change gears nor steer.**

3. Right reduction gear	6 L transmission oil	Remove small inspection bolt.

- **Otherwise** You fill up too much or too little, both conditions are bad.

4. Left reduction gear	6 L transmission oil	Fill up until the oil flows over.
5. Turret drive	5 L transmission oil	Fill up until level is one finger's width under fill plug hole.

- **Otherwise You cannot swing the turret around!**

6. Blower drive	7 L transmission oil	Only up to upper level with the engine turned off.

- **Otherwise** the oil is thrown onto the exhaust manifold cover.

Oil level:

Too much oil is just as bad as too little! With the engine running at 1000 rpm and warmed up to 50°C coolant temperature measure the oil level and fill up to the proper level. Repeat the process after travelling 5 km, if possible.

- **Otherwise** the amount of oil in the engine will be incorrect.

Do not lose oil:

Check that the seals are in excellent shape. Tighten the fill- and drainplugs. Follow along all lines to find leaks. Check the oil for signs of foaming and oil slung out at radial seals. Clean the bottom of the sump, drain through the bottom plug.

- **Otherwise** the Tiger will burn.

Oil change:

Change the oil before and after the winter, also within specified intervals. Change it especially after performing repairs on the engine.

- **Otherwise** a new engine must be installed.

In the winter time:

You can run the military engine oil labeled (winter) without a problem down to temperatures of -30°C. Below 30°C You must drain 4 Liters of engine oil while the engine is warm to the touch. Substitute 4 Liters of gasoline for the oil drained. Mix oil and gasoline by running the engine at fast idle for a short time.

- **Otherwise** You will freeze up.

After driving for three hours:

The gasoline in the oil will have evaporated as long as the engine was above -60°C and warm. You can keep moving but must replenish the 4 Liters of gasoline with the engine running, before shutting it off. Whether or not all the gasoline has evaporated can be checked with the bubble meter.
The transmission oil for the military is good up to minus 40°C, therefore it need not be thinned out.

- **Otherwise** the next morning the engine will be frozen up.

28

Motto: Only with his juices under full pressure will the Tiger show his strength to full measure.

Oil Pressure

The correct amount of oil alone is not sufficient. Oil in the pan is just as useless as beer in the cellar, if there is no pressure to pump it upstairs to the hot and dry throats of the consumers. Only then will the place start swinging. Only then can You change gears up and down with a clatter, throw up the dust and swing Your turret like a flag in the wind.

The oil pressure gauge:
With the engine running, the gauge must show a pressure of at least 3 atmospheres. When on the move 7 atmospheres is the right pressure. If a line bursts or gets clogged or if the bearing clearance has become too large, the pressure will fall. In that case You must immediately shut off the engine.

— **Otherwise** the engine will freeze up.

The oil filter for the engine oil must be cleaned with each oil change or, even better, more frequently.

1. Remove lid, remove filter pack.

2. Loosen the wing nut, remove the filter plates and separator plates one by one.

3. The housing and the separator plates must be cleaned in gasoline. Caution! The gasoline contains lead and will damage the skin.

4. To install, first slide one filter plate, then in alteration one separator- and one filter plate over the suction pipe. Then install the top plate and press the assembly in place with the wingnut.

5. Install the filter pack. Do not forget the top pressure spring!

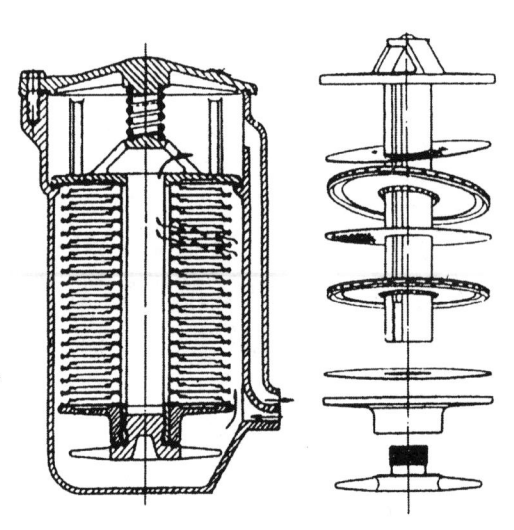

Moral:
As generally true in life, we note here:
Only pressure gives the right atmosphere!

Before the race, a runner will run around the track a few times in order to warm up. If he starts cold he may tear tendons, but he will not break records.

Waiting

Before beginning to move, the Tiger's driver will let his engine idle, in the summer for 5 minutes, in the winter for 15 minutes, until the engine coolant temperature reaches 50°C, the transmission is warm to the touch and the oil pressure has risen to 3 atmospheres.
 - **Otherwise** the bearings will be shot!

Run at fast idle so that the engine turns between 1000 and 1500 rpm. Do not remain at base idle.
 - **Otherwise** the spark plugs foul.

Do not lazily wait around, wait on the Tiger!

Regular service procedures: Lubrication

M O R A L :

You'll find the fittings,
even with much dirt attached,
when looking
at the lube chart with despatch.

Motto: The movie star, the fur she'll grease, the driver pays more attention to the chassis.

Those who take care of themselves are superior to others. Rather more often and thoroughly should the creme for day and night be applied.
 - **Otherwise** You'll get in trouble with Your supply sergeant.

Motto: Air will give – by virtue of compression the proper bang upon digestion.

Dust is Your enemy!

If You go a distance of 7 km, Your wide tracks will throw up the dust from 1 hectare of land. You will be recognized from far away and will lose Your most effective weapon – surprise.

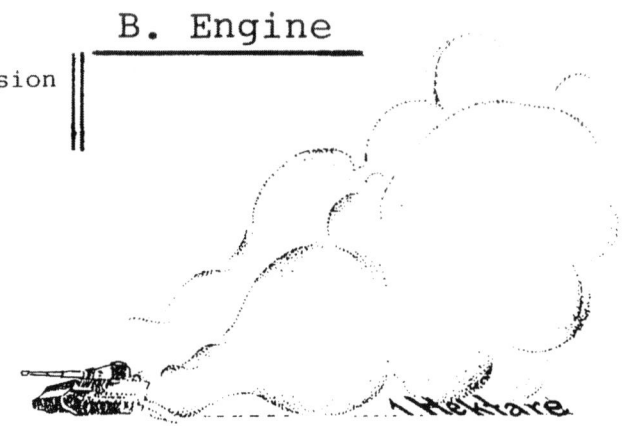

B. Engine

Dust is Your arch enemy!

As You are going these 7 kilometers, Your Tiger uses 170,000 Liters of the same dirty air through which You are holding Your breath.

Within 15 minutes it must swallow as much dust as You would breathe in during a period of ten days spent riding on the back of the tank, where the air contains the highest amount of dust.

Both of Your filters must digest all that dusty air. They are Your only weapons against this deadly enemy.

The air filter catches dust like a flycatcher catches flies. But as soon as it is covered or saturated completely, it is no longer any use. The air then enters the cylinders almost without having been filtered at all, the fine dust is ground between cylinder walls and pistons, constantly working like sand paper. With increasing wear the consumption of gasoline and oil rises because the pistons are slapping inside the cylinders.

In addition, a saturated filter does not allow enough air to go through. The engine now draws in an increasing amount of gasoline, which in turn washes the lubricating oil off the cylinder walls. For the second time wear is increased together with fuel consumption, this time due to a lack of lubrication.

Both factors multiply each other, soon causing a breakdown. A new engine must be installed.

In action Your Maybach engine will go 5000 km easy, if You give it clean air to breathe. Otherwise it will not even go a distance of 500 km.

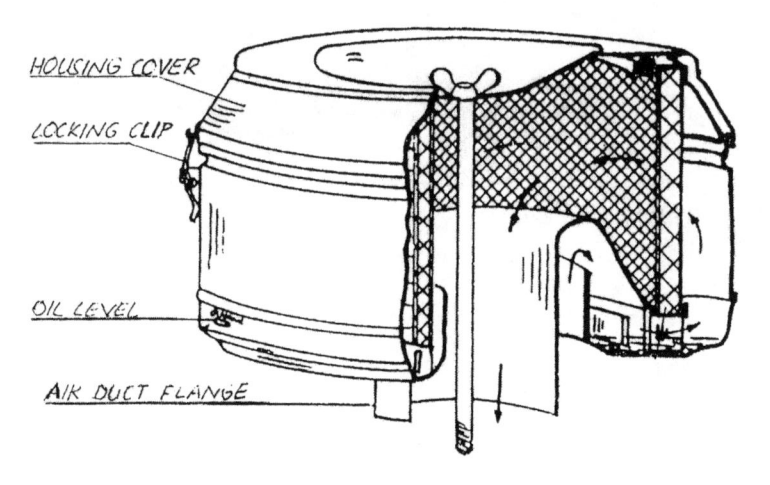

Labels on diagram: HOUSING COVER, LOCKING CLIP, OIL LEVEL, AIR DUCT FLANGE

Therefore:
Clean the air filter after covering any distance which involved any generation of dust!
Loosen the wingnut, remove the filter from the intake air duct, take the assembly off the tank. Remove the lid, then remove the insert. Wash the filter and the housing in gasoline. (Caution, poison!). Dry thoroughly afterwards. Fill up used engine oil to the red mark. Install the insert, watch for a good seal. Clamp on the lid. Install the housing evenly and tightly on the intake air duct and fasten the wingnut, ... do not forget the wire gauze inserts.

They faithfully feed Your engine, but they demand care and attention from You!
Do not drill or scrape with needles or wires, use pliers and a splinter of wood
instead. Do not overtighten the lid!

Clean them often and pay special attention to:

The four
Two Barrel Carburetors

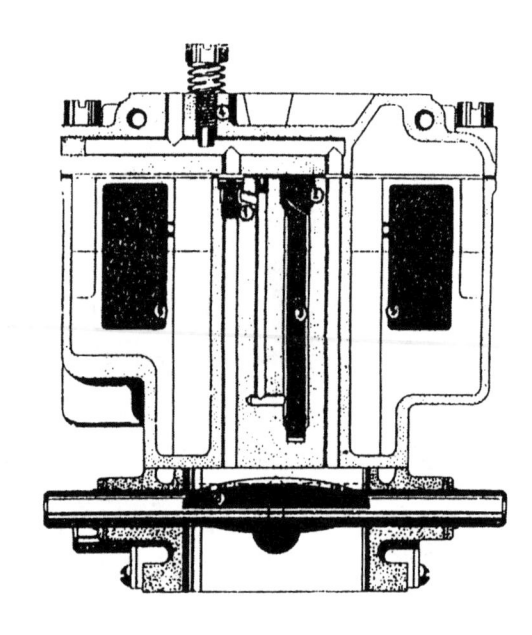

- **The fuel level** (drain by unscrewing main jet).

- **The main venturi**, it must be installed so that "38" or "40" can be read from the top.

- **The center ring**, it must rest squarely on the main venturi. (The center atomizer must not be installed too high ot too low).

- **The throttle valves**, they must close tightly.

- **The floats**, they must not be dented and they must hang free without binding.

- **The linkage**, it must fit into the throttle levers without binding.

- also, that the **side opening** for the idle screw and all **passages** in the carburetor housing are free.
    ...**Otherwise** the engine jerks and backfires.

- **avoid vacuum leaks** by attending to immaculate gaskets and sealing surfaces.
    ...**Otherwise** the engine has no pickup.

Think of the idle speed!
Tunr the air screws in pairs all the way in then turn out until the engine runs smoothly, retain the idle speed setting by fastening the limiter screw on the air pipe.
    ...**Otherwise** the engine will only start with difficulty.

Correct fuel level in the fuel bowl.

Remove the air horn and lay the lower index finger on the edge of the float bowl. The finger tip must get wet when doing this.

    ... **Otherwise,** You will search forever for the defect.

Your guardsman in the engine is the engine speed governor,
- he helps You when the Tiger needs better pickup.
- he warns You, should You be driving carelessly, not watching the oil pressure gauge.
- he checks Your temper when You race the engine.

Because, up to 1900 rpm, You are running on only four carburetors, those of the first stage. The first stage forms the forward part of the carburetors two barrels. It is easily identified by the limiter on the throttle valve.

If the engine speed is increased over 1900 rpm, the second stage is opened by means of the centrifugal governor and oil pressure. The second stage is used for engine speeds between 1900 and 2800 rpm.

They close again as soon as 2800 rpm are exceeded.

If the engine has insufficient oil pressure, a bypass will prevent higher engine speeds - Your sick Tiger must only be driven into the shop.

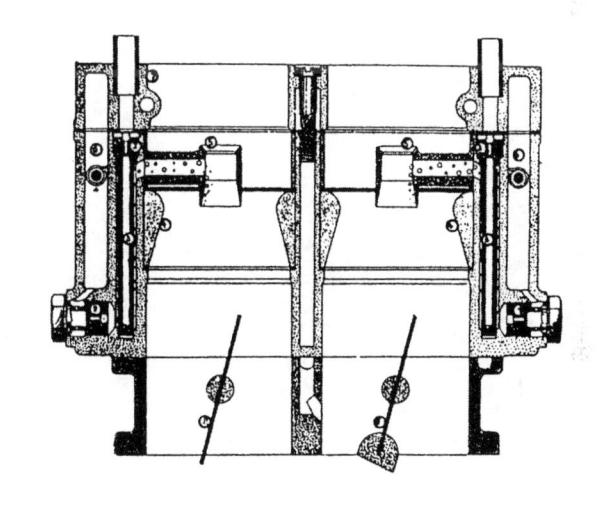

Jet-limerick:
... the wrong idle jet will take revenge,
remember, number sixty five is Your friend!
For the first stage, to learn You should strive
one-five-zero, two-thirty-five:
For the second remember without wonder:
two-twenty-five and two-hundred!

1 main jets (size 235 - 225)
2 idle air jets (size 150 - 200)
3 emulsion tubes
4 main venturi (size 38 - 40)
5 center atomizer
6 float bowl
7 air horn
8 throttle valves

Moral: If the engine jerks and hisses,
it's the carburetor,
that's why it misses!

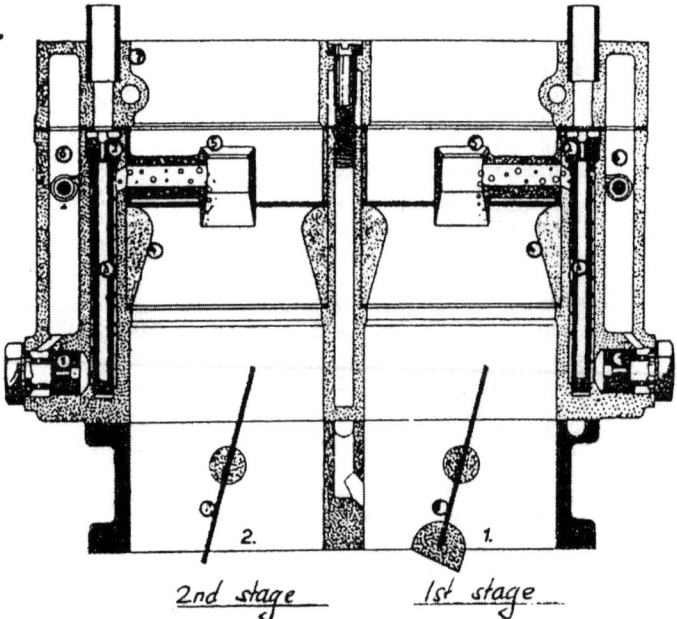

2nd stage   1st stage

35

Motto: Power will only come of applied force, when directed properly, of course.

C. Drive Train

The sliding gear transmission is a thoroughbred race horse. It changes its pace with secure and natural swiftness after very little pressure is applied. You must care for it by the book and keep the linkages properly adjusted. Otherwise it will buck like a fullblood whose reigns are misadjusted and whose leash is not properly hung.

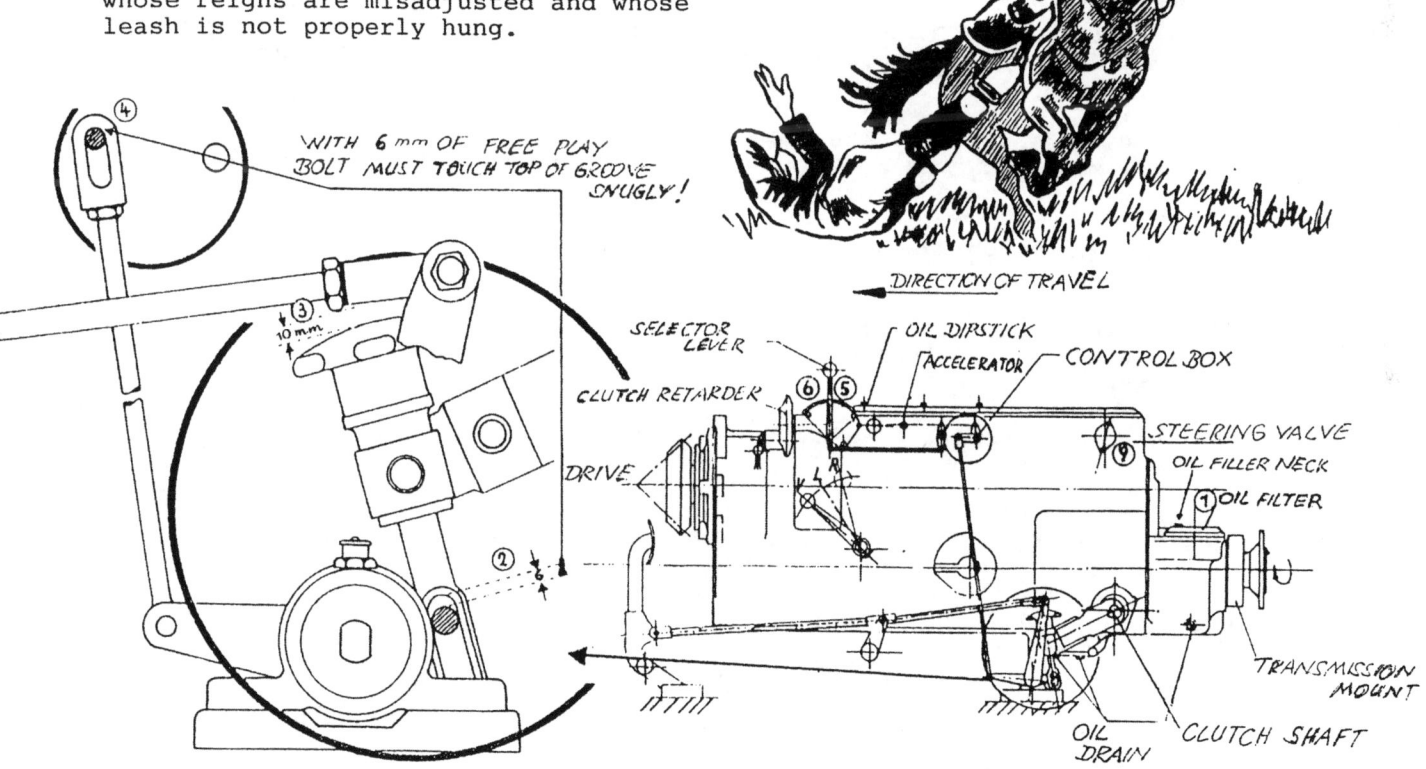

WITH 6 mm OF FREE PLAY BOLT MUST TOUCH TOP OF GROOVE SNUGLY!

DIRECTION OF TRAVEL

- SELECTOR LEVER
- OIL DIPSTICK
- ACCELERATOR
- CONTROL BOX
- CLUTCH RETARDER
- DRIVE
- STEERING VALVE
- OIL FILLER NECK
- OIL FILTER
- TRANSMISSION MOUNT
- CLUTCH SHAFT
- OIL DRAIN

Therefore,

Transmission: 1. Check the oil level, frequently. Clean the oil filter.

2. Turn the wingnut by hand to the right until the clutch is released after a free travel of only 6 mm. Fashion a gauge from wood for the 6 mm measurement.

3. Adjust the limiter on the foot-operated lever, so that the wingnut still travels upwards.

4. The connecting lever to the relay box must be seated without play once the foot-operated clutch lever has traveled through its free play of 6 mm. (See measurement # 2).

5. Adjust the lever on the accelerator shaft so that the engine will reach maximum speed when the accelerator linkage is moved to the wide open throttle position by hand.

6. The linkage on the selector lever must release securely for each gear.

7. Lubricate the linkages and keep them from binding, so that they may swiftly and securely return to the disengaged position.

8. There must always be a little bit of play in the cables to the steering rods.

9. Clean the steering valve, when steering trouble is encountered. The sealing surfaces will be cleaned of any dust particles once the valve plate is pressed in.

10. Retighten the mounting bolts for the sliding gear transmission.

...**Otherwise You cannot shift gears.**

Drive shafts: Tighten the nuts fastening the drive shaft flanges frequently.

...**Otherwise the drive shafts will go their own way.**

The friction surface on the brake itself cannot be renewed. It is glued on, not riveted. You must exchange the entire disc including the friction surface. To do so: Loosen the intermediate shaft and lever, remove the brake from the brake carrier, loosen the screws on the lid together with the brake housing. Readjust them often with a special wrench (21 E 2799 U 15) and replace the radial seal as soon as oil enters through the brake retainer.

>   **... Otherwise they get hot and smoke.**

Auxiliary transmission (steering gear), Check the seals. If oil is being thrown out, they must be replaced soon.

When Your Tiger is traveling 33 km/hour, it has the same thrust as Your armor piercing shell#40 flying at the speed of 3300 km/hour.

If You step on the brake this thrust must be consumed by the friction surface of the brake disc. The Tiger stops after a distance of 12 meters.

If the grenade hits, the armor plate must absorb the whole impact, even 20 cm of steel do not offer sufficient resistance. A braking distance of 20 cm is not enough.

Therefore, the friction surface on the brake must sustain what 20 cm of armor plate cannot. Think of that every time You use the brakes.

Therefore: A free play of 13 mm must be adjusted on the brake. With the brake loose, You can insert a feeler gauge into the inspection cavity. If the free play is above specifications You must readjust the linkage by one further hole.

Motto: One thinks, upon encountering a broken track:
I should have checked, now we must retreat back.

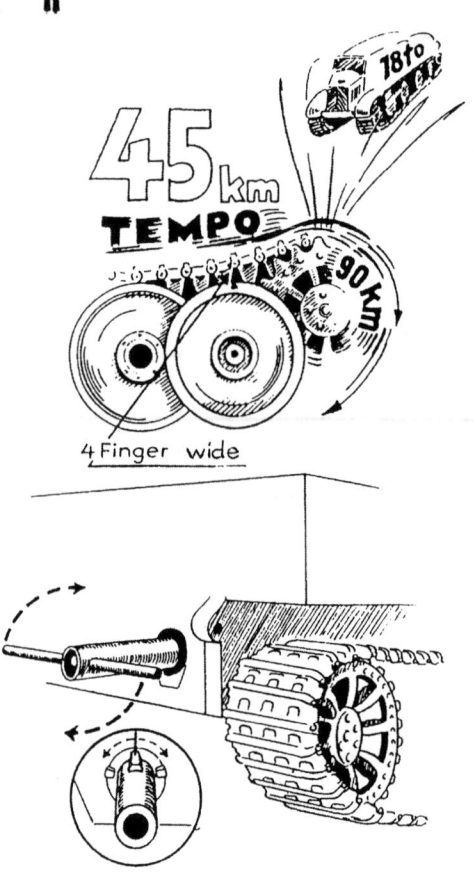

Tension of the tracks is extremely important!
On top, the track runs to the front with twice the vehicle speed, going 45 km/hour, that's 90!
If You do not properly preload the track it will slam onto the drive wheels with a force of 18 tons when steering or braking.
The track should hang four fingers' width over the top of the first roller adjacent to the drive wheel.
When adjusting the preload check the limiter stops and do not overtighten them,
...**Otherwise** the engine must be removed.

Check bolts and nuts on the drive wheel, also on the rollers and the guide wheel. Retighten as necessary. Take care not to damage the sheet-metal locks or replace the locks.
...**Otherwise, the wheels go off by themselves!**

In the winter time all the rollers must turn freely.
Thaw them with a blow torch,
...**Otherwise,** You lose the rubber rims.

Check for loose or broken rubber rims, unlocked bolts, fractured roller discs, broken torsion bars and trailing arms. Exchange them on time.
...**Otherwise the damages will multiply!**

Torsion bars are the joints of the Tiger. You must not injure their polished surface. With them it's like with a love affair. If there is a small fracture, it falls apart quickly. Do not throw any tools onto them, drag heavy or sharp-edged objects on their surface or step on them with spiked boots.

                    ...Otherwise You need to go into the shop.

All terrain track   To install the all-terrain track:

- Completely remove paint, rust, dirt and ice from the flanges.
- Apply a very thin coat of grease,
- Install the rollers.
- Tighten the bolts in a criss-cross pattern and secure.
- Open the transport track under the guide wheel on one side.
- Move the tank forward until the track is off the wheels.
- Lay out the all-terrain track in front of the tank.
- Drive the tank forward until the end of the track is close to the first roller.
- Tie a rope around the drive wheel threefold.
- Hook the rope into the track.
- Block the other drive wheel using the steering lever.
- Pull the track onto the rollers.
- Lock the track links and put tension on it.

The other side is installed in the same manner.

D. Running Gear

Transport track:

The transport track is installed in the same manner as the all-terrain track. In this way the rollers can be removed easily, because they hang freely.

Bolts and track links are replaced with the weak link under the drive wheel or the guide wheel.

New links must not be replaced close together, but distributed evenly over the length of the track.

Change the sprockets on the drive wheels as soon as the forward edge of the teeth has worn off.

This is no centepede, but a Tiger from underneath:

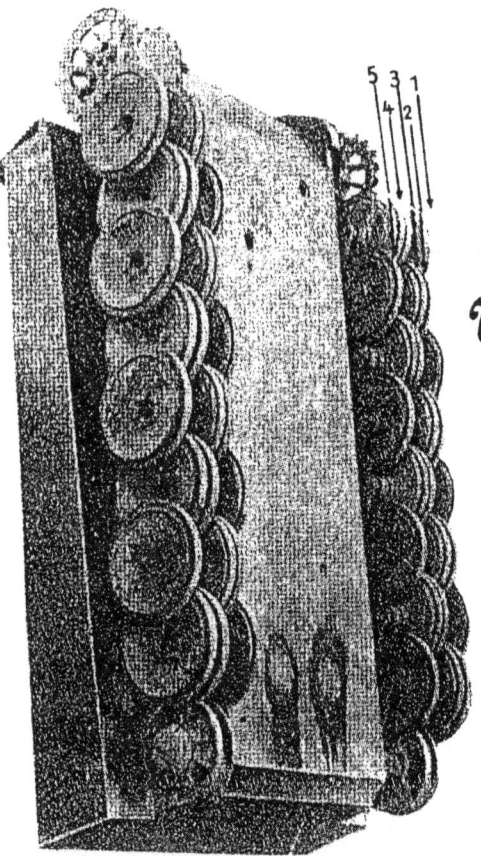

Moral: When it's dark like inside a cow's bud, cold, wet and dirty, full of mud,

jacks and winches stuck in the ground hammer and wrenches nowhere to be found

When bars break, crank arms drag three rollers missing, five of them snag,

one contemplates upon such disaster, the engineer, is he the master?

This is an assessment of the tasks, wrenches and special tools involved in order to change a roller, a drive wheel, a guide wheel or a flange:

Roller in row	1	2	3	4	5
How do I lift the trailing arms?	Lay ramp in front of innermost roller of the arm to be lifted, drive tank onto ramp, (1) Jack up one side of the tank (2) using two solid support bases and two bottle jacks of 30 ton rating.			Open the track, lift one side of the tank over teeth height, using winches.	
How many rollers must be removed?	1	3	4	8	13
Which special tools and sockets are needed?	27	27	27	10 (2799/5) 70 50 C 2798 U5 see (4) stud M39x1.5 bolt 18x35	15 (2799/5) 70 50 C 2798 U5 see (5) stud M39x1.5 bolt 18x110
How many rollers must be removed?	1	3	3	5	
	Outer flange	Inner flange	Guide wheel	Drive wheel	
Which sockets, special tools are needed?	27	27 2798 U10 screw- driver see (3)	22 50 C2798 U5 bolts M14x90 studs M39x1.5 pipe of 15mm ID 75mm long see (6)	Wrench size 50, 46 Remove drive wheel using screws to press off, remove piston with spring. Device C2798 U3 with spindle and nut, socket size 27 and 46, machine screw 50, remove link, remove split ring, replace felt washer, see (7)	

42

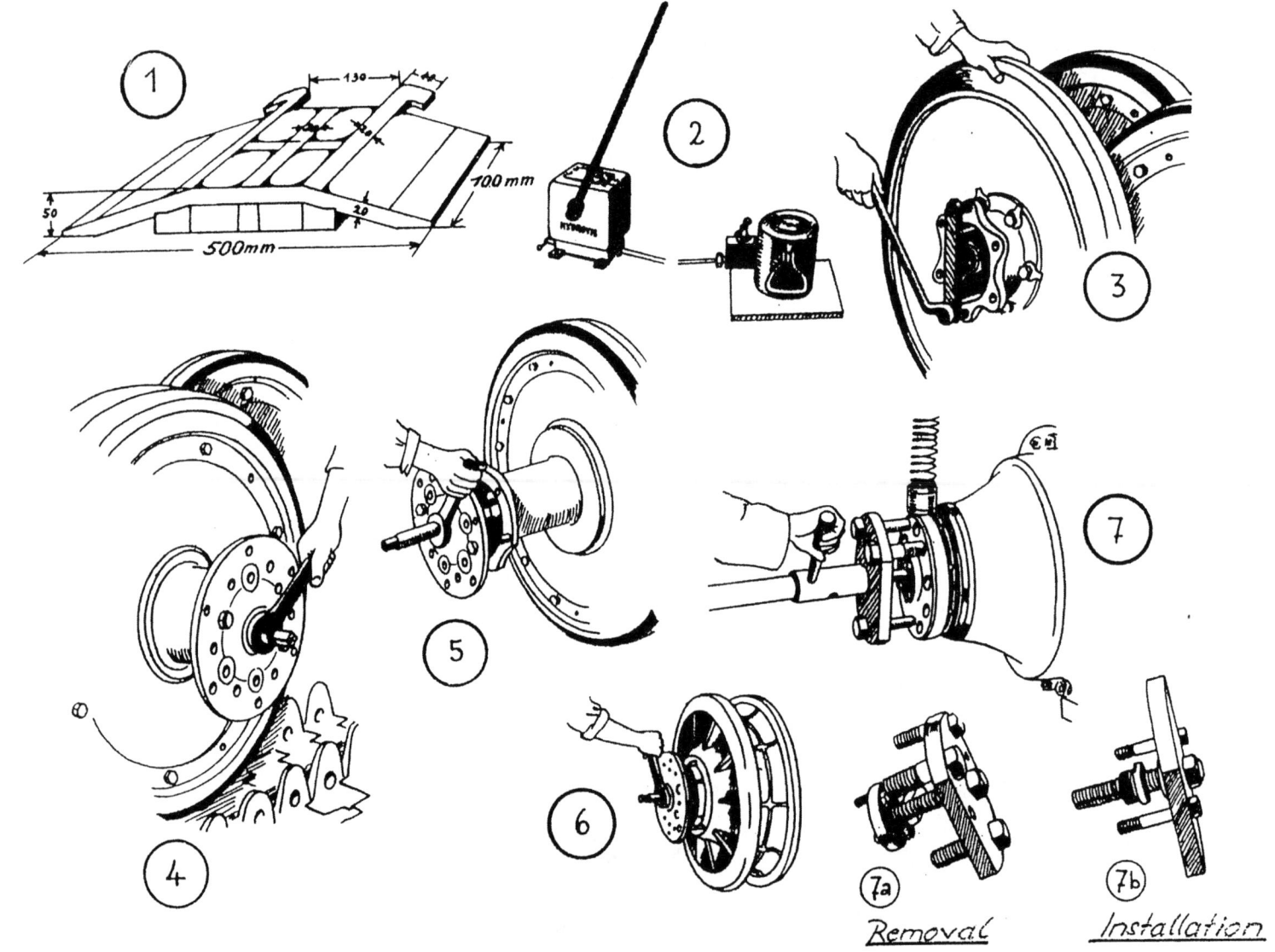

1. 130 | 50 | 100 mm | 500mm | 20

2.

3.

4.

5.

6.

7.

7a. Removal

7b. Installation

Defensive Driving

MOTTO: The Tiger is, one might reflect
a vehicle that handles perfect.

26 turns a minute in the three quarter step is what a fine gentleman will do dancing the waltz. At this pace the music melts in Your ear and harmonizes with the regularity of motion. Going slower is boring, but if You turn too fast You will get dizzy and Your partner will melt for all the heat.

2600 rotations per minute in the four cycle is what the Tiger loves. At this speed he will perform best for the fuel consumed. Your instinct, Your ears and Your tachometer will tell You when You have brought Your partner into the right heat.

Do not race her over 3000 rpm ever, otherwise she will overheat. The water boils, the oil ceases to lubricate, the bearings, pistons and valves burn and freeze - the dance is over...

Therefore drive with Your head, not with Your bud!

Constantly check the speed (1)	coolant temperature (2) and oil pressure (3).
Find the best way to go	but hold the direction,
Approach cautiously	but keep on moving,
Check what is ahead	but read the gauges.
Report on the intercom	but listen to engine and transmission.

On the move — Turn the cannon to the 6 o'clock position and tie it down.

Buildings and walls — **should not be run over!** The rubble looks better in the weekly movietone news than on the tail end of Your Tiger. The blower will suck in all the dirt and dust, the radiator is covered up and no longer functions. The engine overheats and fails.

Tarp, leaves, rubble, luggage — Must not lay on the blower cover or disturb the cannon when rotating the turret.

Morass, swamps	Avoid dark areas and high grass. Prefer to make long detours. Investigate the ground on foot. Take another man piggyback and stand on one leg. If the ground carries You, it will carry the tank. Go through swiftly, do not steer or change gears. If You get stuck, do not dig Yourself in attempting to get out. Another Tiger will pull You out. Anchor the cable, hook into the tracks and pull Yourself out.
Log dam	The dam must be 3.5 meters in width and the logs must be at least 15 cm in diameter. Otherwise they will break or work loose when passing over the dam.
Rivers	A solid riverbed and firm riverbanks are necessary. Where other tanks wade through the water, the Tiger can go too. Turn off the engine and prepare for underwater driving. Close the sump vent, turn on the bilge pump.
Bridges	Investigate on foot. Prefer to ford. Stop in front of the bridge. Position the Tiger so it can be crossed without any need to steer. Select low gear, do not change gears, do not stop, drive slower than walking pace. Accelerate only after 5 meters of having crossed the bridge.
Ditches and craters	Approach head on, avoid wet areas.
Wooded areas	The Tiger will tear down trees up to 80 cm in diameter using the edge of the front plate. If the clearance between trees is too narrow, drive in a zig-zag pattern, with one side running free.
Mines	Stay on the tracks, bump back on tracks, do not steer, eliminate mines if possible.
Snow	New dry snow is no reason for concern below 70 cm in height. Compacted snow or sleet only up to the level of ground clearance - 50 cm.
Ice	Throw chainlink in front of the track, use inertia, do not steer, approach edges or ditches with one track. Using twigs or sand for traction makes little sense.

This is Your confession chair.
You need to know Your way around
here, so You'll find all the
levers and switches even in the
darkest of nights, like at home
the light switch, the doorhandle
or,...well You know what.

Driver's lookout shield
must be kept movable! In the
wintertime and when under attack
it can sometimes jam. Loosen the
four countersunk screws in the
frame. Remove the lid and turn
the eccentric bushings far
enough to the left, until there
is enough play in the
adjustment knobs.

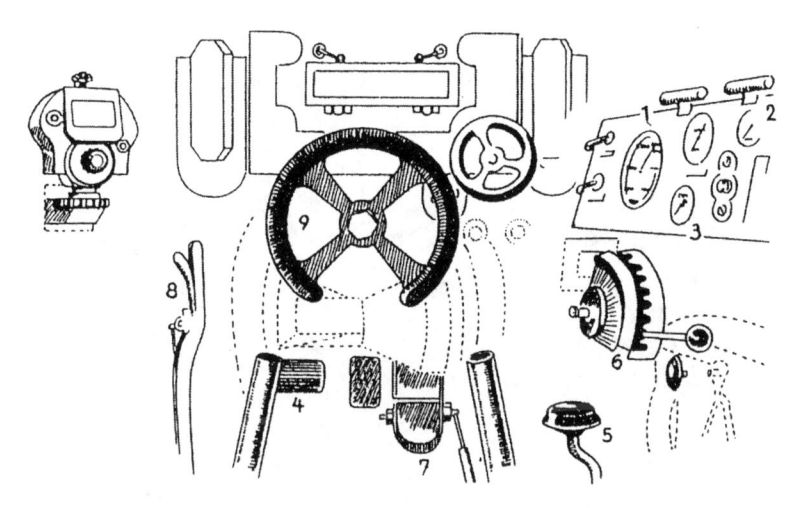

To start out:
1st - 4th gear
(5th - 8th gear
impossible)

4. Depress clutch pedal
5. Directional lever forward
6. Selector to 1 - 4 detent
   Engage the selector
7. Accelerate,
   slowly engage the clutch.

One gear is always engaged.
If that one is suitable to start
out, You will not need to change
the selector.

Upshift:
8 gears

Selector in detent
Engage selector lever

No need to reduce engine speed
or use the clutch. 1 or 2 gears
can be skipped once the transmission is warmed up. Watch the
tachometer!

Downshift:

8. Selector in detent
   Adjust the brake lever
   with feel,
   Engage the selector

No need to use the clutch,
no need to double-clutch or
speed up the engine.
1 to 2 gears can be skipped
once the transmission is
warmed up. Watch the tachometer!

In turns	9. Shift down before the turn. Pull in by the larger or smaller radius according to feel. Using any given gear, a wide or a narrow turn can be made. The smaller the turn, the smaller the gear must be. If it doesn't work out, brake lever, change gears.
Turning on the spot	Shift down to first through third gear, depress the clutch, pull left or right. Push the large button on the transmission housing.
Stopping	Shift down to first through fourth gear, Brake lever, Depress clutch, Directional lever to "0", Engage clutch.
Backing up, 4 gears	Depress clutch, Reverse directional lever, Selector lever to detent. Engage selector lever. Accelerate engine and slowly engage the clutch.

The directional lever cannot be moved to "0" or reverse as long as a gear above #4 is engaged. If You stopped while in 5th through 8th gear - depress the clutch - shift down.
Reverse is only possible in 1st through 4th gear.

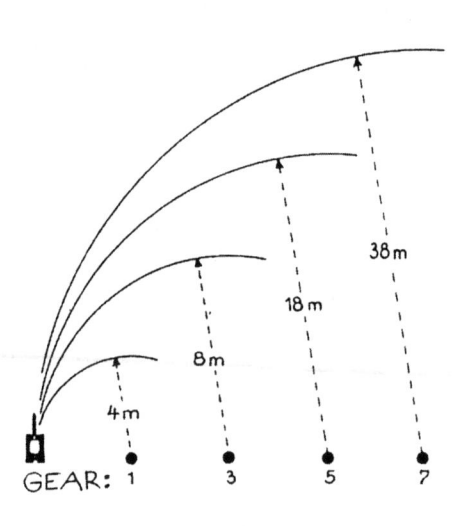

GEAR: 1 — 4m
3 — 8m
5 — 18m
7 — 38m

Order to shoot	Depress clutch, apply hand brake lever
	Try out position 10½ and 1½ o'clock and memorize. Commander and gunner give directions via intercom.
"In position"	Engage selector lever in second gear.
	The position of the three shafts and the respective gear can be noted on the plaque on the transmission housing.
"Breakfaaast"	Pull rrright, or...
"Luuunch" (see "Daily meals")	Pull leeeft
(see "Eyeballing")	
	Look out - estimate distance - report - look out.
Emergency shift	Directional lever to "0" Use a wrench to change gears Depress the clutch, Directional lever forward. Accelerate, Engage the clutch.

Moral:

‖ As with all things which one may but not must, driving is a pleasure, full of lust. ‖

Engine Shutoff

Carbonic Acid (CO_2) is refreshing. What sparkles in lemonade, foams in beer, tingles in champagne, is carbonic acid. It rejuvenates smells sour and tastes great, as everyone knows it will get You drunk.

Carbon monoxide (CO) is deadly. The exhaust gases contain carbon monoxide apart from carbon dioxide (CO_2), the former being a profoundly evil substance. You do not see it, You do not taste it, You do not smell it. You will slowly get tired, unconscious, pant for a few minutes and drop dead!

*Sparky Innocent rests here
he died of tragic atmosphere
in his Tiger it was cold
Lo and behold ...
So he let it run warm
but the smoke was not forlorn
since a tarp against the rain
blocked the pipe whence it came
So the fumes slowly crept
to Innocent who just slept
Five more times he'd breathe
'til upwards his soul took leave
Surely, if still awake,
the tarp off the exhaust he'd take*

Ventilation is the only means of protection against it. Carbon monoxide is heavier than air and slowly settles in lower areas. You must use that condition to Your advantage!

Therefore:
When turning off the engine, open both sides of the engine cover, open the sump vent, open the lookout, open windows, doors, close both fuel cocks. Remove ignition key. If the engine does not shut off - accelerate to wide open throttle position and turn the main battery switch to "0".
- **Otherwise the Tiger bursts!**

In the wintertime do not stop on the barren ground. Lay twigs, brushwork, logs, straw or fences underneath. Remove dirt, mud and ice between the rollers. Move the tank a bit every two hours.
A sudden change between warm weather (thawing) and cold weather (freezing) is especially dangerous.
- **Otherwise it will freeze in place!**

Thin the oil, see "6 x Check Oil", and remove the batteries. If stopped for a longer period of time, see "Power". Engage the gear You intend to use when starting out later on. A cold transmission cannot be shifted through the gear range. Clamp down the clutch pedal so the clutch is disengaged and does not freeze onto the flywheel,
- **Otherwise You cannot start out later!**

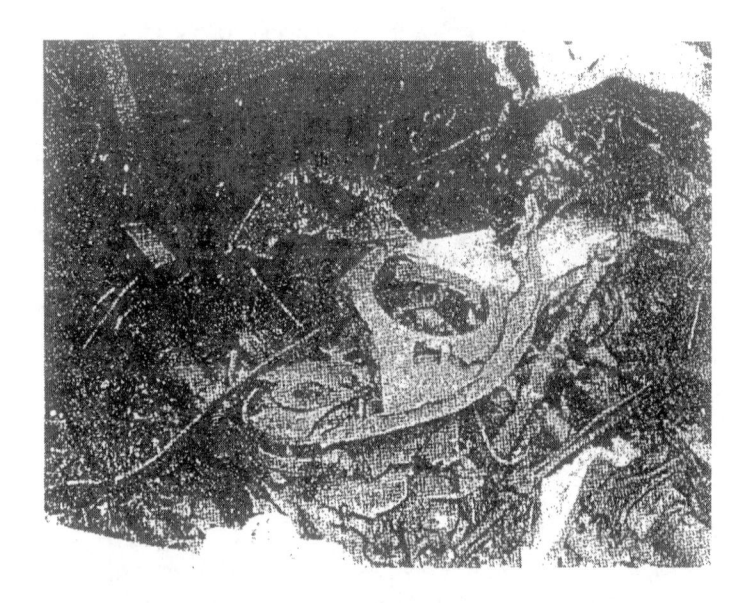

Sometimes it happens real fast:
A leak in the exhaust duct caused carbon monoxide to accumulate in the sump. You think no evil and push the starter button in the morning, and it starts right up, the whole tank that is! A tiny spark from a loosely insulated wire blows up the whole vehicle.

Moral:

‖ The Tiger does not like its own smell, much like the soldier doesn't care for his as well! ‖

Field Recovery

Motto: With care, thought and cool - Recovery is accomplished soon.

Just as You would help Your comrade, no matter what, You must take care of Your friend of steel too and take him home when he breaks down.
If need be another Tiger can help You out, but it is better to avoid that avenue.
It is better to skip any further attempts to get out on Your own. You torment the engine and driveline, and it is no good anyway -

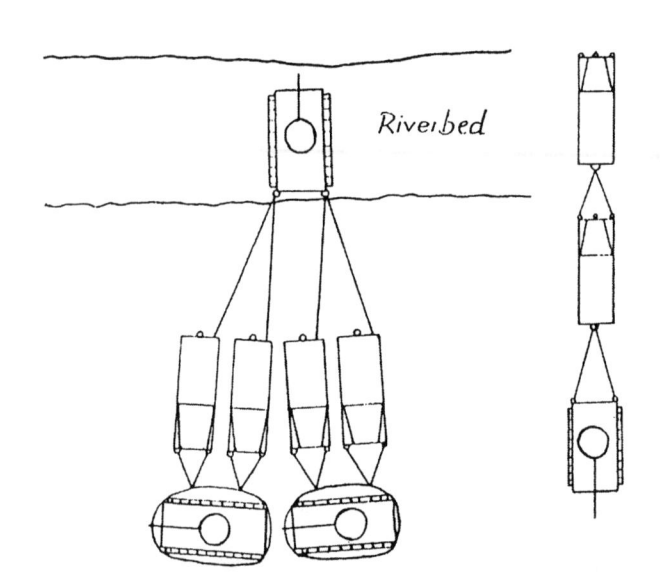

Riverbed

Instead:
Report and let the experts talk! In the meantime, prepare for recovery, paying attention to the following:

Oscar:
frees up the tracks or opens them to check the running gear,
...so that resistance to towing is eliminated,
- removes the steering gear box shaft and replaces the bolts,
...so that the transmission is disabled, but the brakes work.

Barrelbum and Weenie:
remove obstacles in front of the tracks and hull,
... so that the recovery effort will be less difficult.

Speedy Smart:
has checked for anchor points for the tow tractor and prepares the appropriate tools: breaker bars, tow bars, hooks, ropes and winches,
... in case the recovery will be done using winches.

Don't fiddle around and waste time, or You'll be reprimanded!

Inform the commander of the recovery team on damages to the tank and avenues for recovery right away.

And then everyone lends a hand!

Once the vehicle is free it will be towed in a tandem train.

Be alert as a watchdog when crossing bridges, fording rivers, or passing narrow roadways.

Keep in contact with the towing tractors, make an extra effort giving directions,

Otherwise Your comrades will be broadsided or the tank ends up stuck once again.

Moral:

> Recovery is full of difficulty, yet as much a necessity.

MOTTO : Even General Guderian up the chain of command
    takes the train when in demand!

Loading for Transport

Loading a tank onto the train is smooth and quick business if You have prepared everything properly:

Apply the railcar (SSyms) brake and support the overhang at each end of the railroad car,
- **Otherwise Your Tiger ends up on the rails.**

Install the transport tracks and stow away the accessory rollers, but make sure the track cover is lifted up, so it will not endanger railroad traffic.

When loading the Tiger prefer to use head ramps, lay out both all terrain tracks side by side, drive the Tiger over these, fasten the tracks on the front and pull them onto the railcar in this manner. The remaining ends are folded inward.

Once the tank is on the rail car, do not forget to apply the brakes and chuck the tracks at each end.

While moving by train, frequently check:

- whether brake is firmly applied,
- whether the wooden chucks are still nailed down,
- whether the tank is still centered on the railcar.

MORAL : Loading onto the train,
    for the experienced is just a game!

Radio Operator

"RADIOMAN WEENIE DESCRAMBLED"

    Your set reaches farther than the voice,

    the ear, the eye.

    It travels over distances faster than

    a tank or a projectile.

    The responsibility of whether it turns into

    a powerful and dangerous weapon or into

    a mean traitor is in Your hands.

Motto: Often the proper radio broadcast will divert the attack to the better, at last!

The right wavelength and the proper volume are often decisive for Your future. In turn, a ridiculous mishap, such as a wrong adjustment, a missing connection or a loose contact can ruin everything. Be wired up!

Always:
1. – Plug the wires to the transformer and to the antenna into their proper sockets,
2. – check that all switches are in the "off" position when the apparatus is not in use.
3. – check the connections from the battery, (+ on +, – on –) over connector box 23 in the base plate, and from the transformer to the apparatus for tight contact. Pay attention to loose wires and insulation.

Before using the apparatus:
Connect all wires as shown on the diagram.

To operate the receiver:

Adjust	2	for high volume
Check	4	wether the scale is lit.
and	5	for burning control light
Adjust	6	to "0"
Turn	7	on the ordered frequency and lock.
Adjust	8	for "far away"
Turn	6	to maximum volume
Adjust	8	to "near" if volume is too high.
Turn	2	back if it is still too loud.

To operate the transmitter:

Adjust	2	to position "Tn"
Check	4	if the scale is lit.
and	5	for burning light.
Turn	7	to operating frequency.
Push	9	
Turn	10	until
	11	points to the far right of the scale.
Does	11	oscillate when You speak into the microphone?
Adjust	2	to "Tg sounding" if You want to use the Morse code.

The Radio Apparatus

After operation:
    2   Turn the switch to "0".
    1   and stick the wires into their sockets.

Moral: For two frequencies he plugs his wires,
to hear it all is what he desires.

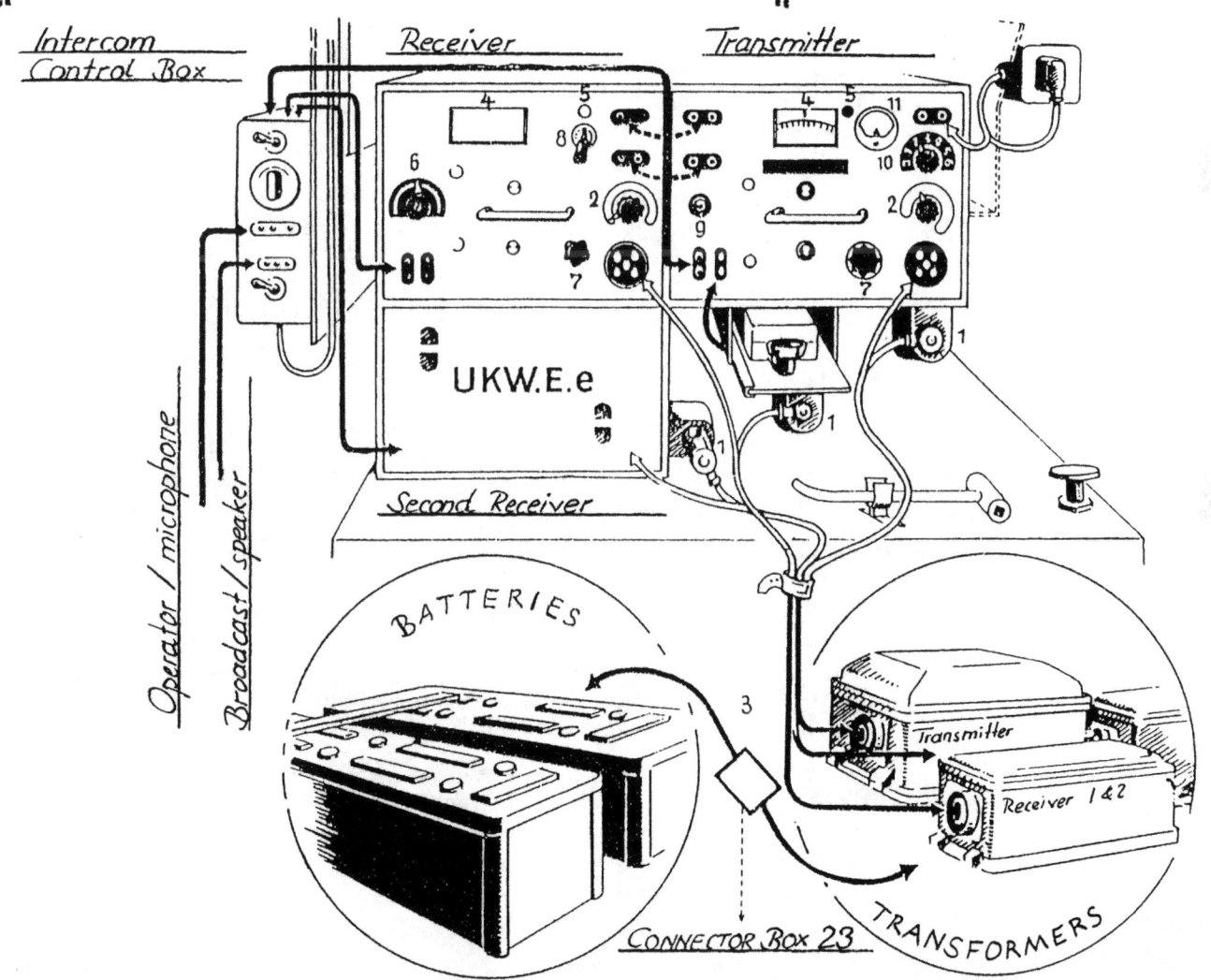

Intercom Control Box

Receiver

Transmitter

UKW.E.e

Second Receiver

Operator / microphone

Broadcast / speaker

BATTERIES

TRANSFORMERS

Transmitter

Receiver 1 & 2

Connector Box 23

Intercom

Motto: Radio and telephone were made to better hear the tone!

This is the intercom control box with its two switches. Using the upper switch You can obtain different settings for the intercom. The lower switch may be set either way. The receiver is turned on, the transmitter is not. If You have no receiver, hook the 5-wire cable from the transformer to the intercom control box.

1. First option: "Intercom"

Tank commander! You can listen and speak without pushing Your button. You must therefore be especially careful, as everything You say will be heard. If You wish to utter maledictions or talk to the infantryman, You either have to remove the microphone or unplug the microphone wire, or have the radio operator turn off the whole apparatus.

If You want to talk to the radioman You must push Your button.

Gunner and Driver! You are constantly listening in. If You want to speak, You must push the button.

Radio operator! You can only talk to the commander after pushing Your button.

2. Second option: "Broadcast and Intercom"

Radio operator! If You constantly want to be connected to the intercom, turn the upper switch to the left, position "Broadcast and Intercom". Like the commander You then hear everything and can speak without pushing Your button.

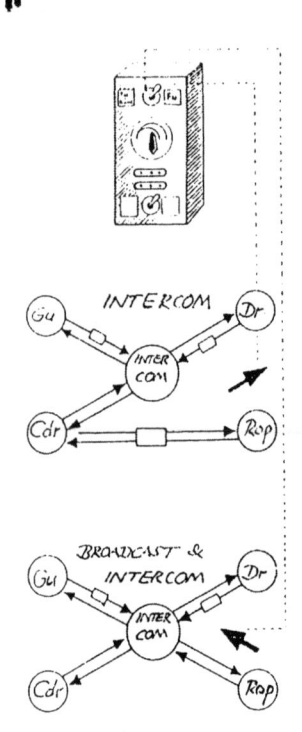

The four speakers of the intercom are shown as circles, transmitting and receiving is denoted by arrows. If the arrow goes through a square then the button must be pushed on the microphone in order to talk.

Moral: On the intercom it works as good, as with a female in early womanhood!

Motto: The broadcast would be beyond description with a female performing the encryption!

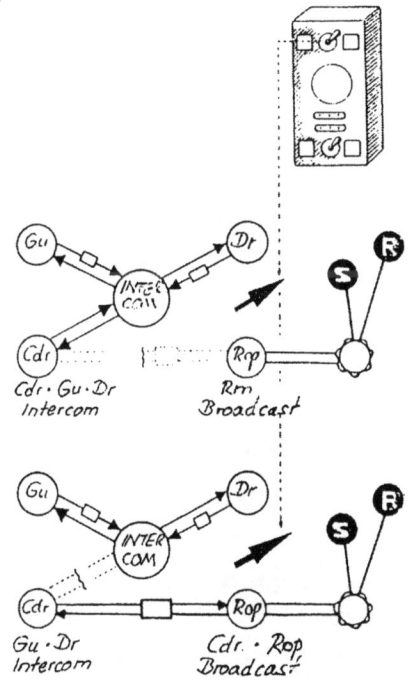

Here 2 x 2 options are possible as the lower switch is now part of the setting. For now, we will leave it set on the right, on:

A. "Commander and Radio operator, Receiver 1 and 2"

1. First option, "Intercom"
Radioman! You can send and receive by turning the operating mode switch to "Tn" or "reception" while the commander, gunner and driver talk on the intercom undisturbed.

Should the commander want to hear the incoming broadcast or if he wants to send out a message, then either You or him need to push the button. You will then hear what is coming in or being transmitted. In the meantime the commander is disconnected from the intercom.

The intercom is denoted together with broadcast mode in the above illustrations.
On top, intercom is to the left and above the dotted line, the radio operator being in broadcast mode.
Above, intercom is to the right and above the dotted line, only between gunner and driver.

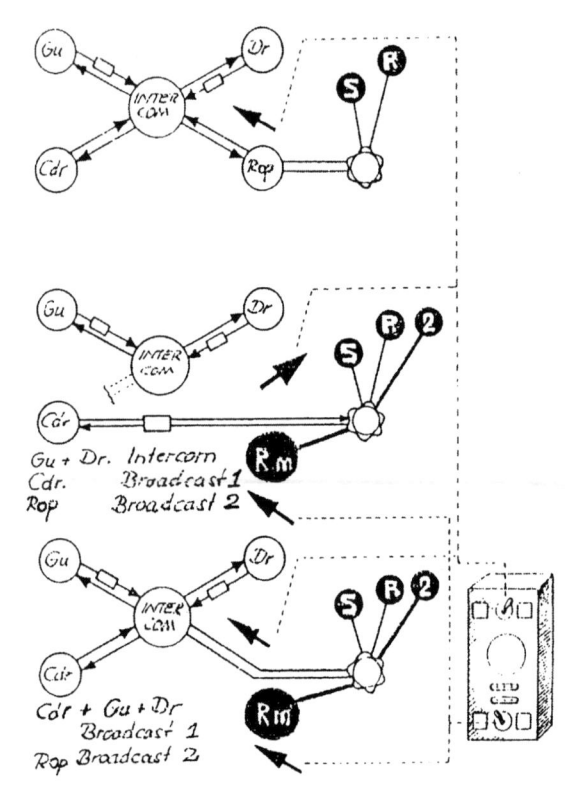

In the illustrations, intercom is denoted as previously. On top only the radio op. is in broadcast mode. In the center one message is denoted by thick, another by thin lines, intercom is above to the right of the dotted line.
Above, thick lines denote message to radio op., the others message to the crew.

2. Second option: "Broadcast and Intercom"
If the whole crew is supposed to receive, switch to the left, position "Broadcast and Intercom".

All four crewmembers are now interconnected, all four can now send. Gunner and driver must push their buttons to do so.
Now special care is to be taken to shut up!

B. "Commander receive 1, Radio op. Receive 2"

1. First option: The upper switch to the right, on "intercom" mode.

In vehicles with type Fu2 and Fu5 apparatus You must listen to both receivers, that does not require a Ph.D. either.

But if two messages arrive simultaneously, then push the button quickly and turn the lower switch to the left, on "Commander receive 1, Radio op. receive 2".

In that case You only keep receiver No.2, while receiver No.1 goes to the commander, or ...

2. Second option: For the whole crew, if the upper switch is turned to the left on "Intercom and Broadcast".

Moral:
Therefore, think fast and act quick, or where You now a lightning bolt will sit

Loader

"BARRELBUM THE FREELOADER"

60 tons of steel and 700 horsepower serve only one purpose, to set in motion and protect the weaponry You operate. If You fail, all of that will be in vain. If You prove Yourself competent, a multitude of enemy tonnage and horsepower will be destroyed with Your aid.

Motto: Often one cannot really fire as she will not as one desires.

DO NOT unwrap too early!
DO NOT stand up but lovingly lay on a blanket.
DO NOT use the packaging material for heating, turn it back in.
DO NOT let moisture, dirt, sunlight or frost touch it!
DO NOT toss or dent like the bricklayers.

Shells with fractures or dents - **throw them out!**
Shells with marred rotating band - **throw them out!**
Shells with leaking explosive - **throw them out!**
Shells without base plate or crimping - **throw out!**

Caution!!!
Inspect, clean, do not lubricate!
Hand tighten loose priming screws!
Priming screws must not protrude!
Tighten nose fuse by hand!
Straighten out loose and rotating projectiles!
Ammunition with percussion primer will cause a short circuit!

Attention!
Insert shells tightly into their mounting brackets!
Rearrange storage on time!
When loading do not marr the rotating band!
Anti-tank grenade #39 is black with white tip!
Anti-tank grenade #40 is black!
HL - grenade is grey!
High explosive shell is yellow!
Only adjust delay using a wrench!
After unloading, set back on O.V., otherwise it will fail!

Turn in duds and used shells!

Moral: Wether blond, black, turning grey, or white, care for her like for Your bride.
Her temper You will come to admire,
With the touch of a button she'll catch fire.

Motto: Jamming in the gun's barrel, thank God, it's rather rare.

The versatile Cannon

Beforehand...
Check Your circuitry, care for the ammunition, clean the lock, rotate/actuate all moving parts, clean and remove all the oil from the barrel before shooting. Apply oil liberally after use when the barrel is again warm to the touch.

Otherwise the cannon will not shoot at all!

Attention...
Remove the muzzle cover, also the disposable one, when covered with ice.
Remove camouflage and brushwork away from the muzzle.
Look through the barrel during a pause in firing.
Shine inside with a flashlight at night.
Remove fragments and residue.
Unload a hot barrel during a pause in firing.

Otherwise the cannon will shoot to the side.

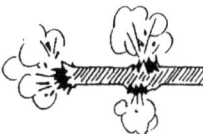

Do not shoot,
if the muzzle brake is loose or shot off -,
it works like a sail and helps to absorb 70% of the recoil.

If the recoil brake loses oil -
it works like a shock absorber and absorbs 25% of the recoil.

If the pneumatic recuperator leaks air or does not function. It works like a door-closing link and absorbs 5% of the cannon's recoil.

If the recoil marker signals "pause in firing".
It must be shifted forward after every shot.

If the cotter pin on the operating lever is loose or missing.

Otherwise the cannon will shoot backwards.

Then again, with experienced marksmen and generally anyway.....

the cannon will shoot forward!

Moral: Only with regret the tanker will admit, instead of scoring does he take a hit!

Motto: The 8.8 is rapidly ignited but some of them never enlightened.

Slow Response

Jamming at...	Cause:	Remedy:
Chamber	Corrosion or dirt on shell	Reload.
Priming screw	Useless (can be recessed)	New priming screw.
Striker	Too short, dull or broken	New striker.
Bridge	Broken spring	New bridge.
Block	Not reached by the bridge	Push cannon forward, refill air to 55 at, (44 L oil)
Socket on pushbutton	Loose wire connection	Repair socket and plug.
Signal lamps	You can fire, even with lamp is burnt or has fallen out.	Install a new lamp, bend the spring as needed.
Oil fuse	Recoil brake is leaking oil (contains 5.1 Liters of oil)	Check seal, tighten screws, fill up oil.
Bosch type plug	Wire pinched, plug is not fully inserted.	Check socket and plug, new wire, bend spring.
15 Amp. fuse	First find the short or pinched wire.	Obtain new fuse from driver.
40 Amp. fuse	Anti-aircraft ammunition, chafed wire	Replace percussion primer with electric igniter.
Batteries	Loose clamps or dirty batteries	Clean, tighten clamps, apply grease.
Remedy for malfunctions up to 15 Amp. fuse:	Lamp on the trigger does not burn, signal lamp does burn!	Switch to emergency battery on emergency switch.
Remedy for malfunctions up to Bosch type plug:	Lamp on trigger burns, signal lamp does not.	Insert wire into socket for turret lights, pull through with the loader fuse.

Check: Unload cannon, hold trigger pulled, lay test light with one end to ground, (bare metal), with the other end on the wire, (insulation removed).

Be careful! Do not cause a short circuit! Check the wiring towards the cannon until the test light no longer lights up. The malfunction is located shortly before that point!

Attention! If the oil fuse has turned power off, shooting must not occur.

Moral: More than one had to bite the dust, rather mud, the wiring, that's what he forgot.

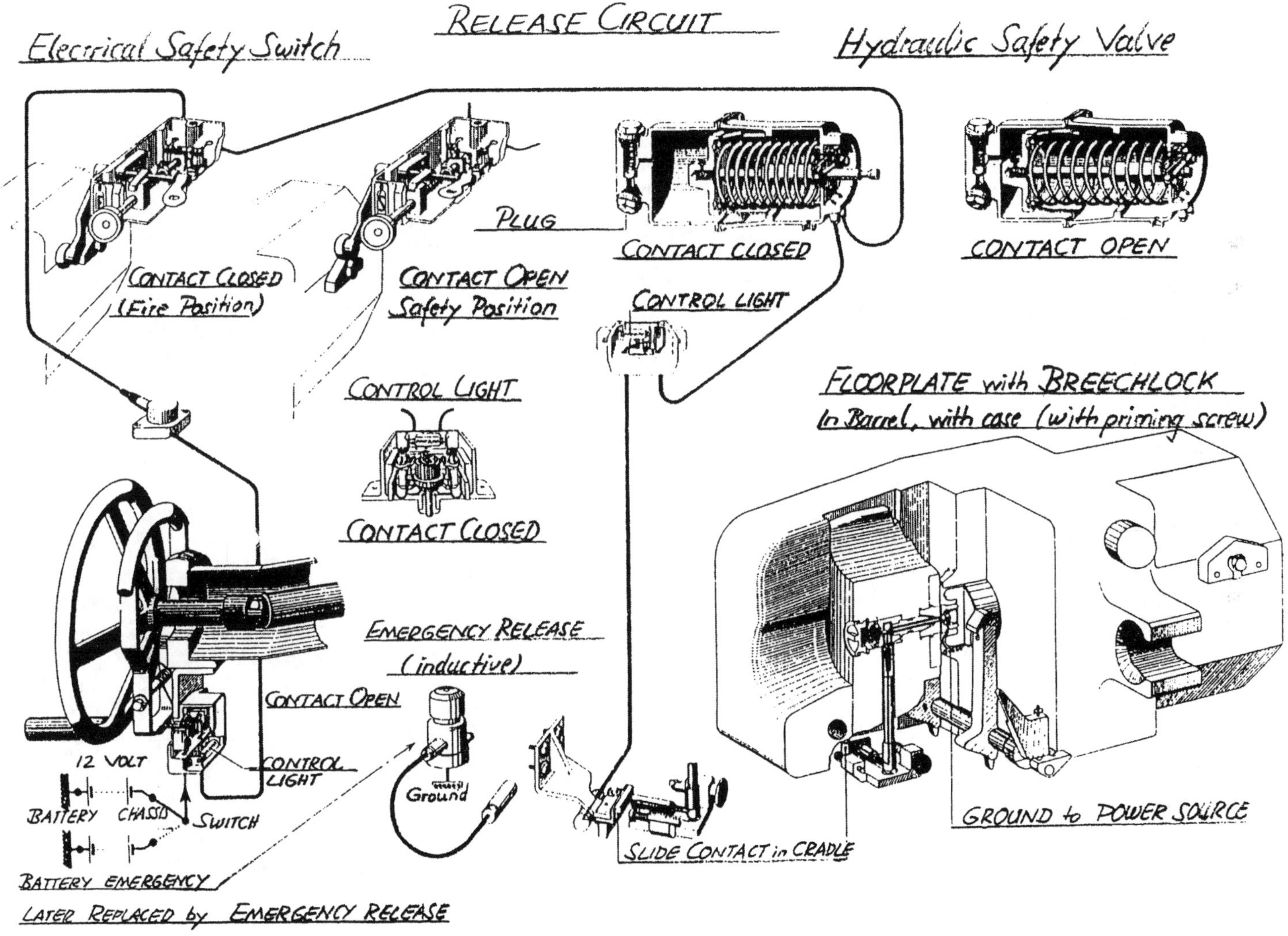
RELEASE CIRCUIT
Electrical Safety Switch
Hydraulic Safety Valve
PLUG
CONTACT CLOSED
CONTACT OPEN
CONTACT CLOSED
(Fire Position)
CONTACT OPEN
Safety Position
CONTROL LIGHT
CONTROL LIGHT
CONTACT CLOSED
FLOORPLATE with BREECHLOCK
In Barrel, with case (with priming screw)
EMERGENCY RELEASE
(inductive)
CONTACT OPEN
12 VOLT
CONTROL LIGHT
Ground
BATTERY
CHASSIS
SWITCH
BATTERY EMERGENCY
LATER REPLACED by EMERGENCY RELEASE
SLIDE CONTACT in CRADLE
GROUND to POWER SOURCE

Motto: The sloppy one, – when the cannon he must employ, jamming will deprive him of the highest joy!

Bullets:
With dents or fractures, rust or deformation,...**throw out!**
Install only German made ammunition straight out of the package, do not use suspicious Russian ammunition dropped by air (explosive ammunition). Check every bullet, clean, do not lubricate.

5 Cures for Jamming

Belts:
With pockets that were stepped on, are bent or corroded...**throw out!**
With broken or bent claws...**throw them out!**
With links torn off or stepped on...**throw them out!**
With worn off link connectors...**throw them out!**
    Do it like the skiers!
Dip the belts into boiling kerosene, shake off well!
That will last for an average campaign. Install the belts properly, the claw must sit in its groove snugly.
Assemble with care, the stud must be centered in the opening.

Machine gun:
Assemble properly.
Check the length of the recoil spring, (forward to center insert).
Check the length of the firing pin string, three turns over end of bolt. The firing pin nut must snap audibly. Do not insert the belt feed the wrong way.

Oil:
Apply oil only on moving parts and locking cams. Use high sulfur oil or even better, some engine oil. Remove the oil from the barrel and locking cams,   ...**Otherwise** You'll have inhibitions.

Installation:
Proceed so that the machine gun is not distorted. The mounting fork must fit over the pins on the housing without binding. Adjust the trigger linkage with locknut. The machine gun must be set for continuous fire. Move the cocking slide forward, so that the tang does not break. Empty the deflector bag.
   ...**But before installation:**

...Lay Your hand over Your heart and ask five questions:

Question #1: Is the barrel or the jackey bent? Does the counterrecoil mechanism operate?

Check #1: Cock the machine gun, remove the flash damper. The barrel must be easily pushed to the stud using just one finger, but rebound immediately.

Question #2: Does the machine gun operate in continuous fire mode?

Check #2: Kick the trigger, pull the lock and let it snap forward. It must catch only when releasing the trigger, but then immediately.

Question #3: Does the lock operate freely?

Check #3: Remove the base plate with locking spring. The lock must be movable together with the locking slide without effort.

Question #4: Does the lock engage completely?

Check #4: Let the lock snap forward, open the lid. The mating surface of the lower lock housing must be even with the edge of the lower half of the feed mechanism.

Question #5: Does the process of delivery, deflection, extraction and ejection work properly?

Check #5: Insert a few cases with projectiles on top, let the lock snap forward and pull back. The case must be ejected sharply.

New! Swift readiness to fire:
When loading the lock is left in forward position!
You can take Your time loading.
If the safety stop fails, no shot can be fired!
Do not close the lid with the use of force.
If You want to shoot You only need to load through.

Moral:
Check then apart from the belt, does the sprayer work as well?

A Mules's Barometer

Motto: A mule will know from the wiggle of its stick,
whether it's wet out, windy, or hot and thick.
Watching the machine gun's stick,
the gunner finds the jam, that's the trick!

```
If the tail is dry and does not wiggle.........nice weather
If the tail is dry but wiggles........................windy
If the tail is wet but does not wiggle................rain
If the tail is wet and wiggles........................storm
If the tail is nowhere to be seen.....................fog
```

Just as easily, You can determine what the problem is with Your machine gun when it jams:

Pay attention! Remove the foot from the trigger,
on the right side, move the cocking slide back,
while at the same time checking:

1. Position of the lock?
2. What is being ejected?
3. What is in the way of the lock?

Secure on the left,
on the right, remove the lid and check:

And now look at......................

The Machine Gun Barometer

Lock position?	What is ejected?	What jams?	Immediate remedy? Root cause?
forward	cartridge, dud	failure	continue fire (4)
	cartridge, intact	striker	exchange lock (3/4)
		belt binds	pull belt through (2)
	nothing	ejector rod	exchange locks (3/4)
		carrier	pull on belt (3/4)
almost forward	cartridge, intact	distorted gun	loosen claw (2)
		locking catch	exchange barrel (1)
	nothing	cartridge dented	exchange barrel (2)

What is the cartridge doing?

center	cartridge jams, barrel free	improper load	continue fire (2)
		ejector rod	exchange lock (3/4)
	cartridge jams, case in barrel	deflector bag	empty bag (2)
		extractor	exchange lock (4)
		chamber	exchange barrel (1)
	cartridge jams, split case in barrel	loose striker nut	exchange lock/barrel (2)
		bolt stop	exchange barrel (4)
	case jams, cartridge in barrel	ejector	exchange lock (4)
almost back	cartridge not ejected	bent pocket	continue fire (2)
		connector	continue fire (2)
		lock travel	clean (1)
		bent ejector	exchange lock (2)
	cartridge travel	belt travel	continue fire (2)
		feed, upper part	continue fire (3/4)

All the way back, caught after 1st. shot:

		linkage short	pull off by hand (2)
lock does not stay in place	(if it is to stop, hold the belt)	linkage binds	pull off by hand (2)
		trigger dirty	reload often (1)
		wear on trigger	get other machine gun (4)

Root cause: YOUR FAULT   NOT YOUR FAULT

1. Dirt! Clean, remove oil apply oil and graphite.
2. Sloppiness! Reload the belt, straighten out, readjust.
3. Fatigue! Sagging springs, extend the springs.
4. Fracture and wear! New part from spare parts bin or ordnance shop.

Moral: You see, if the shots don't go as they came, usually it's You who is to blame.

EN-RA-DRI-LI-CLEAR magic recipee

Motto: llleft - slllow,
rrright - swwwift!

Exterior: **EN**gine cover closed, engage lock
RAdio operator's hatch closed
DRIver's hatch closed
LIghts removed
CLEAR shot

Interior: Gunner,   disengage lock

Driver,   start engine

1. Loader,   engage rotating gear
llleft - slllow,
rrright - swwwift!

2. Radio op.   Selector lever on "turret"

3. Loader,   Emergency lever on transmission upwards

4. Gunner,   Rotate, by stepping on pedal.

reeear - leeeft,
frooont - riiight!

Gunner,   aim using elevating and traversing mechanism.

Driver,   accelerate, when it has to happen quickly.

Moral: Reeear - leeeft,
frooont - riiight!

Motto: Still at home, for some there went the turret and their head out of detent.

Turret Trouble

OR...

Moral: With sense we swivel elegant— who has none labors by hand.

Trouble:	Cause:	Remedy:
Turret cannot be rotated by foot	Clutch stuck	5 Leave engine running, knock clutch loose!
	Clutch linkage too short or too long	6 Loosen nut on clutch linkage and adjust fork!
	Linkage jumped out of ball pivot	7 Connect ball pivot and secure!
	Center shaft dislodged on top of slip ring connector	8 Remove bell housing and engage center shaft dogs!
	No oil pressure	Refill to proper level!
	pedal off hinges	9 Install linkage, new cotterpin
Turret hangs up at 4 or 8 o'clock position	Turret hanging up on open engine cover	Using a rope attached to the cannon, pull to 12 o'clock position, close engine cover!
Turret swings only to the right when operated by foot	Spring under pedal is too long	4 Adjust spring or set pedal in horizontal position!
Turret swings at different speeds left or right, when operated by foot	Pedal linkage too short or too long	9 Shorten or lengthen pedal linkage!
Turret swings without interruption	Clutch and linkage binding	6 Turn off engine and free up linkage!
	Compression spring stop not seated right	5 Remove drive shaft at flange, unscrew the locknut, pull off the clutch, do not damage the needle bearing, insert spring stop straight, install clutch.
Emergency lever does not function	Emergency lever is turning with shaft, pin sheared off	3 Install new pin to secure emergency lever!

Gunner

"GUNNER GLASSEYE, ALWAYS ON TARGET"

Aiming a shot into dead center is
a matter of art, but not black magic.
In order to shoot better than Your
opponent, You have been given the
sharper weapon and the sharper mind.
Using the 8.8 You can shoot
off a mosquito's right canine tooth.
Here You learn how:

Motto: Never will You learn to aim or shoot,
if You haven't eaten Your way through this book!

Barrelbum had received a gigantic cake from his bride Elvira on the occasion of his birthday. The cake had a diameter of 2 kilometers.

This cake was to be shared with every man in the division, so Barrelbum cut it into 6400 pieces.

those were mighty strange pieces of cake. If You started eating at the tip, there was hardly anything to bite, because it was so thin. But further back it became much wider, all the way up to 1 meter in width at the outer fringe. Each piece of cake was 1000 m long.

Elvira would have liked to bake a cake where each piece would have been 2000 m long, but the field post declined submission to the unit. Those would have been 2 m wide at the end.

You can easily figure out the width for such a piece of cake, if You only know the distance from Your mouth:

             For 1000 m, it is 1   m wide
             For 2000 m, it is 2   m wide
             for 800 m, it is 0.8 m wide
                                          and so on...

The really smart ones will say the width is always 1/1000 or one thousandth of the distance from Your mouth.
         Such a piece of cake is called a notch!

4 notch for instance is as wide as 4 pieces of cake side by side. Look out! That is where the reticules in Your scope are located!

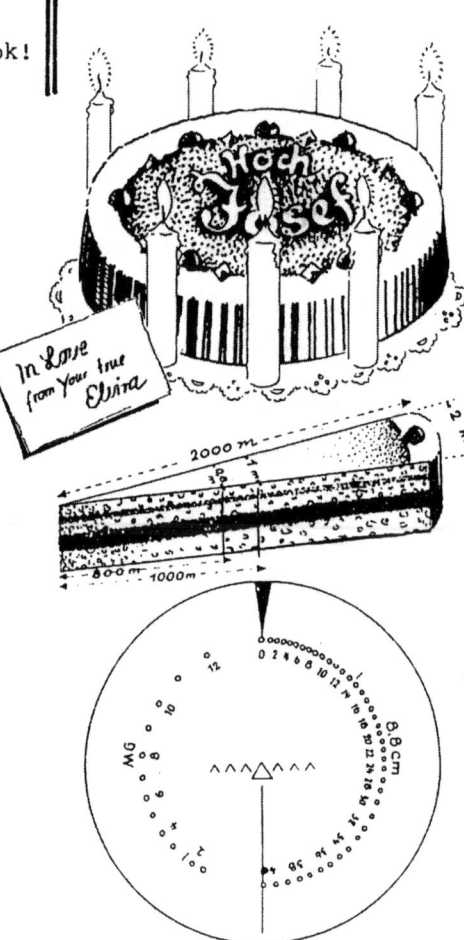

The Notch

The tips of each set of reticules are exactly four notch apart,
If You bear over ahead of them, it is the same as if You looked along the edges of Your piece of cake. So if there is a house 2000 m distant, which fits right between the tips of two reticules, You know: "Look at here"!

    The gap between two tips is   4 notch
    One notch at 2000 m is 2m wide   4 notch x 2 = 8 m
    The house is 8 m wide   Isn't that a killer?

Question: 1 tank is 500 m distant, it reaches across
         3 gaps between reticule tips. How wide is the tank?
You calculate:
        3 gaps at 4 notch each   = 12 notch
        1 notch at 500 m is 0.5 m   12 notch x 0.5 = 6 m.
Answer: The tank is 6 m wide.

You can calculate the height of the target in the same manner, if You know: The center reticule is 4 notch high, the side reticules are each 2 notch high. You must remember this well. You need it constantly when using the scope.

Question: How tall is the tank?
You figure: Assume it is 3 times as tall as a minor reticule,
         3 times 2 notch, that is ... You take it from here.

The real smart ones know that such a scale is also found in the sights of binoculars and may be used in the same manner! But it is also in Your thumb! Hold it far away and it is exactly 40 notch wide. One jump of the thumb measures 100 notch. (Close one eye, then the other, while looking along the same edge of Your thumb). That way You determine the target size and distance in between to an accuracy of 5 notch, stunning everyone! Try it! So if You know the distance, You can calculate the size of the target!

Moral: Are You smarter by one notch? Don't read further if not!

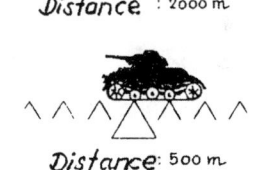

Eyeballing

Motto: Your sweetheart likes it close to see Your eyes,
when eyeballing we separate close and distant, very wise.

Estimating a distance exactly - cannot be done.
"Measuring" is learned by many -
Adjusting the range correctly - must be learned by all!

When estimating a distance of 1200 m or less You must learn to be off by no more than 200 m up or down. If 500 m is the correct distance, then Your estimate must be between 300 m and 700 m. That really is not much of a challenge. Above 1200 m estimating turns into guesswork.

Estimate closer by!
- For dark targets
- If it's dreary and cloudy
- In windy and foggy air
- Against a dark background
- If the sun or reflections hit Your eye.

Estimate farther away!
- For lighter targets
- In fresh and sunny weather
- In clear air without wind
- Against a light background
- With the sunlight across the plain
- Through the scope if You cannot see what is between You and Your target.

Estimate twice: 1. The target is surely closer than X m (e.g. 900 m)
2. The target is surely farther away than Y m (500 m)
Take the average of the two estimates, in this case 700 m.

Distance can only be estimated by the **driver** or the **commander**, since they have a clear line of sight. It cannot be done well through the scope because 1. the scope magnifies 2½ fold, and
2. You cannot possibly estimate with one eye.
Close one eye and let a comrade hold his finger ½ m in front of You. Then try to quickly grab it with Your index finger.
Attention: do not use Your finger and do not look with both eyes beforehand. However, the **gunner** and the **commander** can **"measure"** using the scope and the optics. You will now learn to do that.

If You have the time, do it like this:
    THE TANK COMMANDER
Measures or estimates his distance,   see "Measuring"
    THE DRIVER
(he needs a little longer) reports   see
his distance   "Eyeballing"
    THE TANK COMMANDER
calculates the average   see 1. grade
    THE GUNNER   in preschool
(In the meantime having measured
or estimated for himself) reports   see "Order
his distance   to shoot"
    THE GUNNER
(distance is not the correct range)   see "Belly
adjusts the correct range   Button Rule"

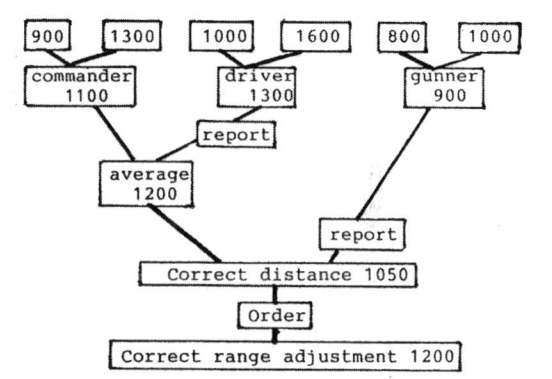

You always have the time!
    If You miss it will take far longer, cost
    more ammunition, You give away Your position
    before Your action takes any effect.

3 times 2 eyes see more than two - You estimate +/- 100 m
3 times, the commander must calculate - he gets more money
3 times, reports and orders are exchanged - that is the
purpose of the intercom.

Practice does it all!

Attention! The correct distance is not the correct range.

```
  900  1300      1000  1600      800  1000
    \  /           \  /           \  /
  commander       driver         gunner
    1100           1300            900
         \         report
          \       /
          average
           1200
                          \
                          report
                             \
                  Correct distance 1050
                             |
                           Order
                             |
                  Correct range adjustment 1200
```

|| ***Moral:*** Closer, foggy, dreary dark, mist, moving, against the sun, Bright, with the sun, clear, plain, between, optics, -farther! ||

Measuring

MOTTO: Even artists measure as viable because eyeballing is not reliable.

If the painter wants to accurately measure a line he will compare the size of the pencil with his model. You compare the size of the reticule with the target! Because if You know the size of Your target, You can use the notch to figure out how far away it is.

Look out: The Russian tanks are all 3 m wide. Assume that it is at a distance that lets it cover 1½ gaps between the tips of reticules. Then You say:

Look at here! 1½ gaps at 4 notch each = 6 notch
6 notch = 3 m
1 notch = 3 : 6 = 0.5 m
0.5 m x 1000 = 500 m

If the tank is positioned at an angle, You cannot calculate using length or width, You use height. Let us assume that the optics display a picture as shown in the illustration. Then You figure:

3 minor reticules 2 notch high = 6 notch
6 notch = 3 m, etc.

Through the scope the graduation looks like this:

Problem: Calculate the distance of this truck!

A few measurements:

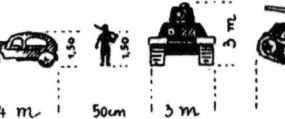

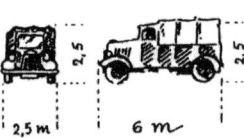

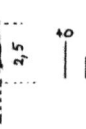

4 m	50 cm	3 m	6 m	2,5 m	6 m

Attention! The correct distance is not the correct range!   **MORAL:**

Moral: Instead of measuring how far off measure by meters and the notch the meters by the notch You divide times 1000, You're on the safe side!

Seven Goodies

Motto: Save ammunition, do not waste – a wagon full is enough for the "Ritterkreuz"!

Pistol:	through the hatch at guests on the rear.
Sub.-M.G.:	through the hatch into ditches and nests in obstructed areas.
Pineapples:	through the hatch into holes and at hidden targets.
Smokeshells:	in case of fire, jamming, in need of tactful retreat, if things get hot and stink.
M.G., front:	as far as 200 m at man, horse and wagon.
M.G., turret:	as far as 400 m at man, horse and wagon, (farther if several present), set fire to buildings, help the infantry by nailing the enemy on the ground.
Cannon:	**High Explosive shell:**
	(no delay) Generates shrapnel 20 m to each side and 10 m forward. Therefore, better to miss to the side than to the rear of the target. Tried and tested against antitank guns, howitzers, amassed targets, nests. Destroys armor, wheels, tracks, lookouts.
	(delay) A mine, if delivered vertically: Intrudes and detonates wooden bunkers, buildings, shelters, forest and juvenile tanks. Incinerates everything and topples vehicles.
	(ricochet) Upon impact after shallow trajectory it will bounce back off firm ground and detonate 50 m further away at a height of 4 - 8 m in midair. Use against invisible positions which cannot be fired on otherwise.
	Armor-piercing shell, #39: Cracks tanks and trenches as far as 2000 m.

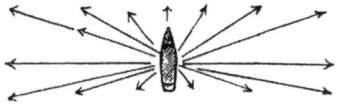

Armor-piercing shell, #40:
Cracks heaviest tanks as far as 1500 m (deviation).
Use only when #39 does not penetrate. Attention!
There is more thrust behind it! **For 600 to 1000 m
You must decrease the range by 100 m, for 1100 m
to 1500 m always by 200 m.**
HL-shell:
Against heaviest tanks as far as 1000 m (substantial
deviation). Blasts enormous holes, but travels slowly.
Therefore the **range must always be increased by $\frac{1}{4}$
compared to the rule.** (for instance not 600 m but
750 m). Do not use if camouflage, brushwork, nets
are in front of the target. Otherwise it will detonate
too early!

Attention: The correct distance is not the correct range!

Moral: Shoot less, hit more, keep tight!
"Reichsminister Speer" will take delight!

79

Motto: Many a target can be incalculable as this female is unpredictable!

Elvira gets shot

Rarely used ranges for a target 500 m away 2 m high available in 6 common sizes.

The correct distance is not the correct range

The men of the TIGER didn't want to believe it either. Barrelbum had obtained a circus banner 2 m high with the pretty Elvira on it and posted it 500 m away as a target. That they wanted to plaster, everyone taking a shot at Elvira.

Driver Oscar took range 475, let Elvira sit on the main reticule, took ½ m to the left, like You're supposed to, and fell short - by exactly 25 m.

Radioman Weenie used range 500 and hit the world-famous toes, -precisely.

Then Barrelbum went in, the loader,(having been trained in the third rank), mightily spat into his hands, took range 700, took a deep breath and hit the trigger - boom - the shot went off, right through the much adorned belly button.

Gunner Glasseye shook his head and said that with range 700 the shot should have gone over top of it. Now he went for it all, took range 1000 and hit the head.

Commander Smart took range 1100 and went over it. With that range the magic had ended!

Range 25 m short, no hit! Range 500 m too far, direct hit!!!!!!

The layman marvels, the expert just looks on!

Moral: The right estimate often indeed doesn't yield the hit You need!

Barrelbum always hits

Motto: The old tanker man will desire a hit generated by any fire.

The cannon shoots point blank. The shot therefore goes up to the range adjusted but no further.

If You know the exact distance and fire with range the same as distance, **then You will hit the point of aim.** But You never know the exact distance. If Your guess is short by 25 m, then You will hit the dirt 25 m before the target, just like driver Oscar did.

The 8.8's trajectory is wonderfully stretched out. So You need to elevate the barrel only a little to shoot much farther. You will then still hit Your target closeby with a distant range, if the target is only tall enough. For instance, using range 1000 You will hit all targets within 0 and 1000 m, if they are 2 m high. Isn't that wonderful?

However, shooting at Elvira using range 1000 is still not on the safe side, because if she were only a bit shorter, the shot would go overhead, as it did with Commander Smart.

There are several usable ranges for one target! The smallest of them is the distance, all others lie above that. You can hit Elvira with 6 different ranges.
500 - 600 - 700 - 800 - 900 - 1000.
Do not adjust range equal to distance! Because if You are short by 25 m in Your estimate, You will miss by 25 m. Do it like Barrelbum, be moderate, then You'll hit the center of the target, the belly button.

In estimating the distance he can afford a glitch of 200 m either way, and he will still hit. Barrelbum always hits, because he can't be off by more in estimating a distance.

Moral: The optics old foxes will adjust further than by estimate they must!

Barrelbum's Belly Button Rule

Motto: $V_{bellybutton} = E + \dfrac{\frac{H}{2}}{E} \times 100$
$$\phantom{V_{bellybutton} = E +} 1000$$

That is the only thing You do not need to remember.

A. If Elvira were twice as tall, then twice as many ranges could be used. Belly-button range would then be 1000. You can be off either way by 500 m!

B. If the target is very small, as for instance the toes, then only one range (500) will do, the exact distance: Anti-tank positions, tanks behind a hill, trenches, weakspots on tanks, as for instance the turret (so that the shot will hit vertically), must be fought in this mannner. Your estimate must not be off.

C. If Elvira moves further away, then ever fewer ranges will suffice.

D. In the end there is only one course left to take: Range equals distance.

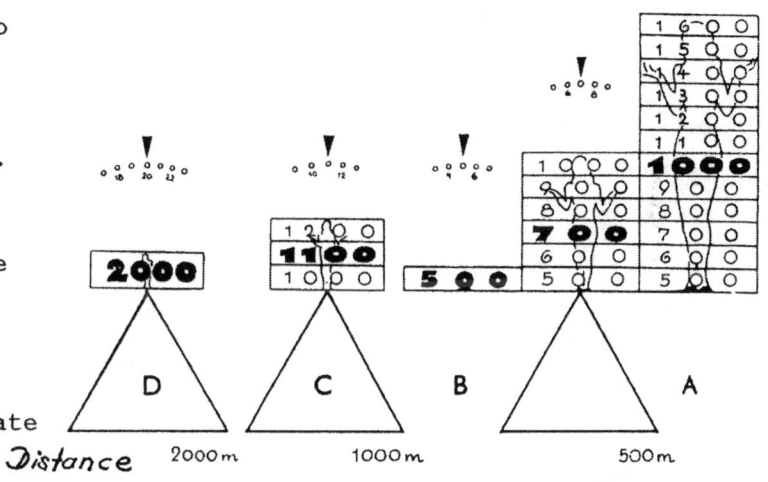

If **the target is very small**, or if it only appears small because it is far away, then the number of useable ranges is also **small** since the target is only a few or not even one notch tall.
Only **small** errors in estimating are permitted.

If **the target is large** or if it looks **large** because it is close by, the number of useable ranges is also **large**.
Large errors are permitted.

Distance: D 2000m, C 1000m, B —, A 500m

How do I find the correct range?

1. **Estimate the distance.**

2. **Estimate the height of half the target,** (bellybutton), in terms of notch by comparison with the reticule (or take the height of the entire target and divide it by two).

3. **Half the target in notch times 100 metres,** add that to the distance, that gives You the belly button range and You will hit the belly button.

By how much can I be off in the estimate?
    Notch by 100 metres,
        is how far You may be over or under in estimating the correct distance and still hit.

Example:

1. Less than 600 m
   more than 400 m, average = 500 m
2. Target height is   4 notch
   belly button thus at 2 notch

3. Permissible error in estimate
   2 times 100 m .... = 200 m
   therefore all estimates between
   500 m + 200 m........... = 700 m
   500 m - 200 m........... = 300 m
   are applicable.

Moral:

The reticule compare
to the belly button where You stare,

to the distance You add
notch by 100, not so bad,

100 m times the notch
is how far You can be off.

𝔐𝔬𝔱𝔱𝔬: Like Max Schmeling his right hand holds,
save the grenades as the battle unfolds.

Always hold below the target, take aim from underneath.
        Attention!
The cannon always fires ½ m, the machine gun 1 m to the
side. Because the cannon is positioned ½ m, the machine
gun 1 m to the right of the optics.

Therefore always hold the cannon ½ m,
              the machine gun 1 m **to the left!**

Under 1200 m
You can't possibly miss, when correctly using the
belly button rule.

Over 1200 m
most of the time You have to adjust range equal to
distance. Since You guessed very accurately You
will fire,
too close or too far.
Then You must adjust the range, because it was off,
even if only by 50 or 25 m. **Do not alter the point
of aim**, as that makes less of a difference over 1200 m.
Only if the shot misses
to the left or to the right,
are You permitted to change the point of aim sideways.
If that is over 2 notch, then use the minor reticule
to hold the target.
If the first shot is not a hit, You either made a
mistake in estimating or You did not properly adjust
the weapon.
You are at fault, not the cannon.
Up to 2000 m the 8.8 will fire point blank. Only if
firing as far as 3000 m, one of three shots will miss.
At a distance of 4000 m only every forth shot will
produce a hit. (deviation.)

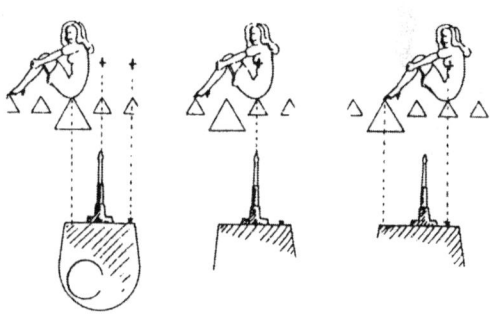

Sensible Use of Ordnance

Therefore, always consider, wether shooting
over great distances is worth it.

After every substantial firing sequence –

Elevate the barrel,
Open the lock,
Let cool and air out,
In the wintertime remove muzzle cover.

Wet the ground in front of the muzzle,
otherwise firing the weapon will generate
dust.

In the winter time, camouflage that spot,
as it will turn black.

Moral: Have the sun from the back,
the wind from the side –
fire from a halt like a crack,
You'll score a hit, great delight.
Yee haw, Yee haw ....

Knife or Fork?

Motto: Whether knife or fork is used on the platter,
You have to eat it, that's what matters!

Some eat with a fork,   others use a knife.

You must be able to use both! Over 1200 m it doesn't always work out right away, especially when using explosive shells. Now the cannon must come to Your aid. It will shoot a scale for You in the field, which You can lay onto Your target just like a yardstick.

Pay attention:
At first always fire one shot with the range 100 m less than the distance You guessed.
Surely that one will fall short.

Moral: Up to twelve by one hundred fire button rule,
farther over knife or fork apply with cool.

If You can watch the terrain behind the target, then use a fork:

For the second shot, add 400 m.
It will go behind the target. Between both locations of impact there now lies an accurately measured distance of 400 m. You must divide it into four equal parts. And now the distance in meters from the first shot to the target can be measured accurately.

The third shot must be right on target!

If You can only see the terrain in front of the target, then apply a knife:

For the second shot, You can add only so much. The location of impact must still be **in front of the target.** Again You have a distance between both locations of impact which can be used to measure how far You are short of the target in meters. Eating with a knife is not that easy.

The third or fourth shot must be right on target!

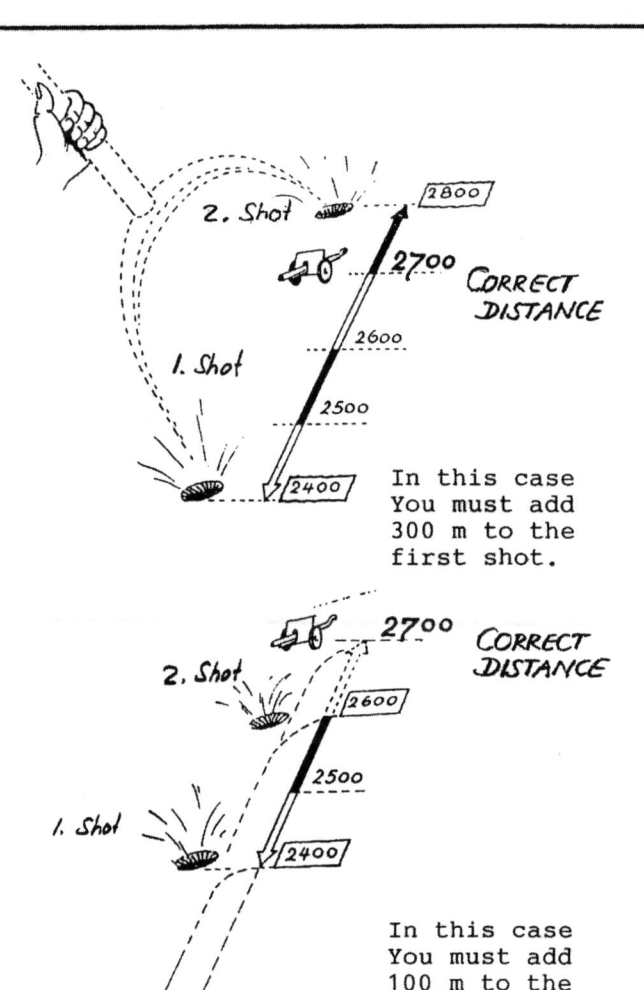

In this case You must add 300 m to the first shot.

In this case You must add 100 m to the second shot.

The Lead

Motto: Easy to hunt whatever's in flight, if one gets the lead just right.

The five men of the TIGER had obtained some cherries while the leave-train had stopped. They now started to spit the stones at the telegraph poles next to the track. That worked just fine. The train slowly started to move. At first they still managed to hit the poles, but all of the sudden they all missed.

Everyone was amazed. Then, Barrelbum shaped his tongue into the barrel of a howitzer, closed one eye, and with the other he aimed a good distance in front of a telegraph pole, pressed real hard, and - boom - the shot went off. Right on the telegraph pole. The faster the train went, the further ahead he aimed in front of a pole.

If Your target is closer than 200 m - aim at it!
If someone wants to cross Your line of sight at a distance between 200 m and 1200 m
                        - **lead ahead of it!**

Because if You aim exactly at him, the guy will have gone a few meters further in the **time Your shell takes to get there.**
It will not hit the spot where he is, but where he was!

First, You must estimate how fast he is going:

	slow, 10 km	average, 20 km	fast, 30 km	
and then take a lead with the main reticule: - for armor-piercing shells 39/40	3	6	9	notch
- for high-explosive shells	4	8	12	notch

Example: Truck is passing straight across at average speed.

"Machine gun 20 shots - 10 o'clock - 600 - truck - take lead 8 notch"!

Always use the minor reticule that holds the target.
That is what they are there for. And always let it run into the main reticule.
If he is not going straight across but coming at You at an angle, then You take half the lead.

Example: Tank is approaching at an angle travelling at average speed.

"Anti-tank 39 - 1 o'clock - 600 - tank - take lead 3 notch!"

If Your target is farther away than 1200 m - stop!
You will waste too much ammunition on moving targets.

The lead measurements are easily remembered by...

Moral: 9 and 6 and 3 - for tanks use we,   12 and 8 and 4 - only for explosive, no chore.

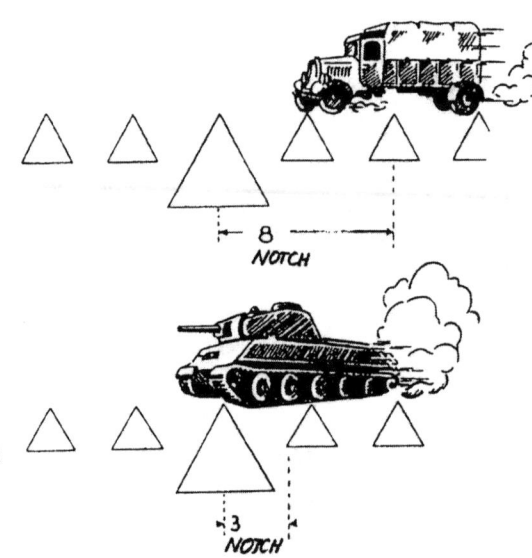

Centering

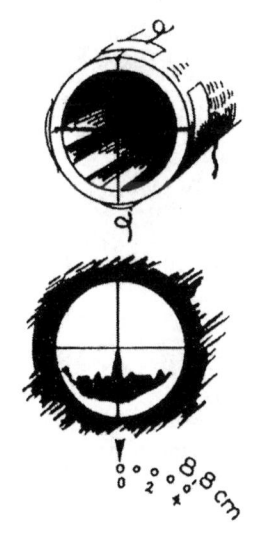

Motto: If You use a fly's specks
when adjusting the optics
and then take aim with great pain
You'll still miss, all in vain!

When on the move always tie down Your gunnery. Still, they will deviate due to vibration. Adjust them Yourself, then You know Your weapon!

First the cannon: To do this You need a piece of thread and electrical tape or grease.
1. Attach a cross hairs across the muzzle.
2. Remove the firing pin.
3. Hold on to a distant target through the barrel.

then the right scope: You will need a square head wrench for the optics.
1. Adjust to obtain a correct focus.
2. Adjust the range for the cannon to 0.
3. Remove the protector caps on the head-piece of the optics.
4. Center the main reticule sideways on the target.
5. Center the main reticule up or down on the target.

then the left scope: You will need a square head wrench for the optics.
1. Adjust the range for the cannon to 1000 m.
2. Hold on to the target with the right scope.
3. Turn the reticule to the left.
4. Adjust to correct focus on the left.
5. Adjust the eye gap until both sights fall into one.
6. Adjust the auxiliary reticule sideways on the target.
7. Adjust the auxiliary reticule up or down on the target.

The emergency range is now fixed at 1000 m. Holding onto a target You can use it to hit any object that is 2 m high at a distance between 0 and 1000 m. Over 1000 m You must go into the target, or let the target vanish.

in the end, the turret - machine gun: You will need a perforated case. Barrelbum always carries one with him.
1. Remove the base plate, take the lock out, insert the case into the barrel.
2. Adjust the machine gun range to the mark between 200 and 300 m.
3. Using the right scope take aim at the target over the main reticule.
4. Center the machine gun through the perforated base plate and the muzzle onto the target.
5. Check by shooting at the target.

machine gun - front: Check by shooting at the target.

Moral: Center Your cannon rather frequently, You'll fire with success, so gleefully!

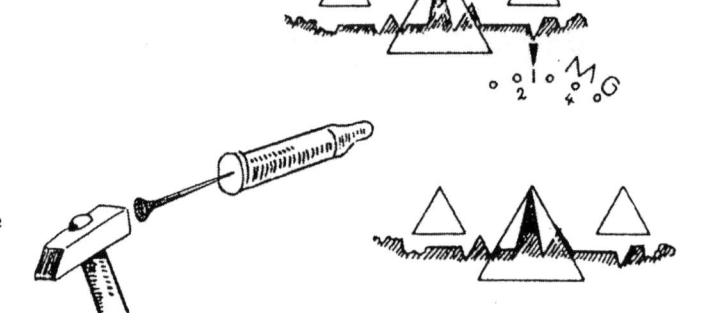

Tank Commander

"TANK COMMANDER SPEEDY SMART"

Only Your clear thought,
Your assured order will
give life to the armor,
direction to speed,
decisive impact to the
projectile.
You hold a hand full of
trump cards, now learn to
play the game.

Order to Shoot

Motto: Since the olden days until today the shot on the target is to stay.

1. Stand musket at an angle in front
2. Stock on left foot
3. Cartridge out of pocket
4. Cartridge onto barrel
5. Take off ramrod
6. Shorten ramrod in front of chest
7. Push in 1 - 2 - 3
8. Wadding rod out of pocket
9. Wadding rod in the mouth
10. Bite off wad
11. Wadding rod onto barrel
12. Ramrod onto that
13. Push in 1 - 2 - 3
14. Feather off the hat
15. Suspend the weapon
16. Clean out the vent hole
17. Feather on the hat
18. Take out powder horn
19. Powder on the pan
20. Powder horn in place
21. Make a grim face
22. Pull the hook
23. Point
24. Aim well
25. Give fire
26. Lord help
27. Fire

In the war of 1618 - 1648, 27 commands were necessary to fire one shot. That is why it took so long. On top of that, the order to shoot was handled differently in various regiments. Some of them used no fewer than 90 commands!

M a k e   i t   s h o r t !
Press Your will into one order consisting of 8 commands!

Motto: 27 main tasks still remain – as shown in this chart – not counting the minor ones! Practice does it all!

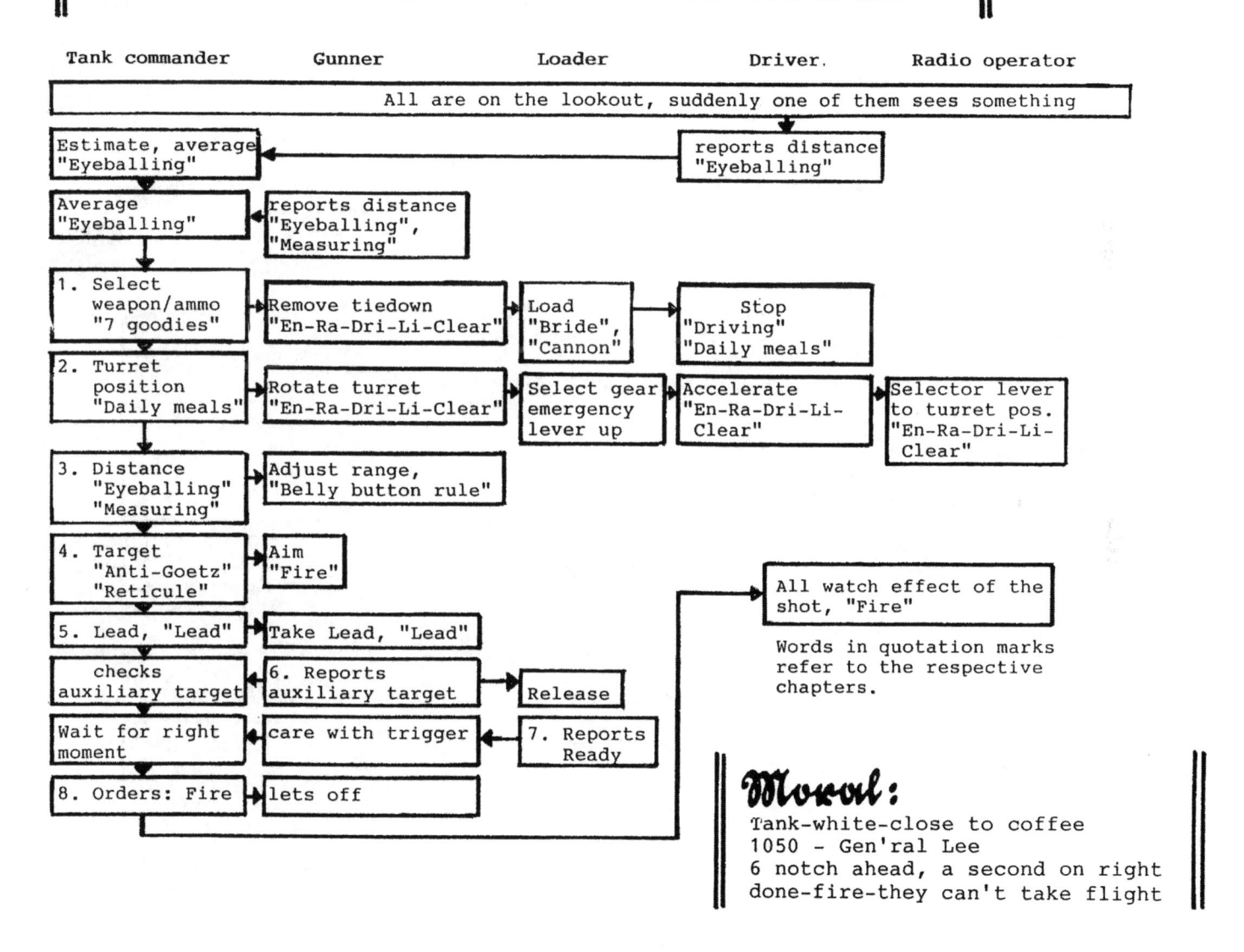

Tank commander	Gunner	Loader	Driver	Radio operator
All are on the lookout, suddenly one of them sees something				
Estimate, average "Eyeballing"			reports distance "Eyeballing"	
Average "Eyeballing"	reports distance "Eyeballing", "Measuring"			
1. Select weapon/ammo "7 goodies"	Remove tiedown "En-Ra-Dri-Li-Clear"	Load "Bride", "Cannon"	Stop "Driving" "Daily meals"	
2. Turret position "Daily meals"	Rotate turret "En-Ra-Dri-Li-Clear"	Select gear emergency lever up	Accelerate "En-Ra-Dri-Li-Clear"	Selector lever to turret pos. "En-Ra-Dri-Li-Clear"
3. Distance "Eyeballing" "Measuring"	Adjust range, "Belly button rule"			
4. Target "Anti-Goetz" "Reticule"	Aim "Fire"			
5. Lead, "Lead"	Take Lead, "Lead"		All watch effect of the shot, "Fire"	
checks auxiliary target	6. Reports auxiliary target	Release		
Wait for right moment	care with trigger	7. Reports Ready		
8. Orders: Fire	lets off			

Words in quotation marks refer to the respective chapters.

Moral:
Tank-white-close to coffee
1050 – Gen'ral Lee
6 notch ahead, a second on right
done-fire-they can't take flight

Motto: Not just the soldier is kept alive by the meal, but also the tank, and that is for real.

Your tank is 12 cm thick at the turret plate
             10 cm thick at the front plate
              8 cm thick at the side and rear plates
Noone else is!

But You Yourself can make it even thicker!
When mother cuts the sausage straight, then that will yield one slice just as wide as the sausage is thick. But if she cuts at an angle, then the slice gets twice as wide!

We're after more sausage now!
If You let someone fire at Your tank straight on, then it is 10 cm thick and will sustain all calibres up to and including 75mm.
But if You stand at an angle, **then it is 13 cm thick.**
A shot that hits at an angle penetrates much less, than one that hits head on. Therefore 13 cm thick plates hit at an angle offer the same protection as armor that is **18 cm** thick against a shot fired head on. (If You want to cut the sausage at an angle You need a sharper knife)

Your armor placed at an angle is therefore in reality as strong as 18 cm and withstands all calibres up to and including 152mm.

Then You cannot be penetrated at all!
You see, just turning Your tank from 12 to 1 o'clock makes it thicker by 2 cm.
In order to penetrate these 2 cm Your adversary has to come 1000 m closer.
1 cm of armor weighs the same as a firing range of 500 m!
If You stand at an angle this has the effect of placing Your adversary 4 km farther distant, in one fell swoop.
From there he can fire all he wants.

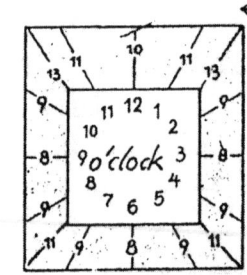

Here You can read all the positions and corresponding armor strenght

... and this shows Your actual armor protection

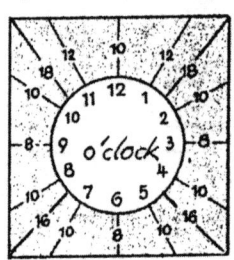

Daily Meals

The best positions towards the enemy are at

$10\frac{1}{2}$ $1\frac{1}{2}$ $4\frac{1}{2}$ and $7\frac{1}{2}$ o'clock

According to the respective hours they are called meals.

To make communication easier, the second syllable is always stretched o u t - (breakfaaast) -

They are easy to learn when compared to the X-mark.

Driver! When taking position always veer to the left or right, until the enemy stands at breakfast or lunch. (Try out the direction and memorize).

Gunner! Fight dangerous targets from the direction of meals at all times. (Read position of turret on the clock and correct the driver).

Tank - Commander!
Approach dangerous enemy at an angle. Order 45° angle position, so that the enemy faces in direction of the meals. (Read position of target on the clock, correct the driver).

Moral: At mealtime - even with a 15.2 -
You might get a scratch or two.
Your adversary has nothing but disgust,
for You, my friend, it's such a blast.

The Cloverleaf

Motto: If enemy steps in this cloverleaf,
You just might be in jeopardy.

At what distance does a T-34 7,62 cm long-barrel
   penetrate my armor?
From direction 12 o'clock under 500 m.
From direction 12½ o'clock under 300 m.
From direction 1 o'clock I am safe.
From direction "Lunch" I am in the safest position.
From direction 2 o'clock under 500 m.
From direction 2½ o'clock under 1300 m.
From direction 3 o'clock under 1500 m.
From direction 3½ o'clock under 1300 m.
From direction 4 o'clock under 500 m.
From direction "Coffee", I am safe, and so on.

If the enemy is located inside the cloverleaf,
I will be penetrated.
If he stands outside, I will be safe.

At "mealtime" the TIGER cannot be cracked.

Men of the TIGER !

>It is in Your own hands, whether the TIGER is
>safe or not. Enjoy Your meals!

Should the enemy really end up inside the
cloverleaf, don't wet Your pants right a way.
Instead, turn the TIGER towards "mealtime".
Immediately the other guy is outside of it
again. If two are firing at You simultaneously,
turn one onto "mealtime" and blast the other one.

We look at the TIGER
from up above

lay our watch around it

and chart these
distances.

No trespassing into cloverleaf by T 34

If done for all hours,
the lines connecting
the charted dots yield
a cloverleaf.

For an adversary with a longer cannon, the cloverleaf is larger.

For enemy weaponry which can only penetrate less, it only has three leaves, because the front is then safe at any distance.

Only one number
    is what You need to memorize for each enemy tank.
    Then You will know the exact size of Your cloverleaf!
    For the T-34 with a 7.62 long-barrel it is.............

1500 m is the length of the three long leaves!
    (because the TIGER is equally thick on the side and arrears).
    Always 1000 m shorter than the longer leaves is the short one.
500 m, (because the TIGER is 2 cm thicker in front).

The really smart ones
    can also calculate how far they may let the enemy approach them for positions 2, 4, 5, 7, 8 and 10 o'clock, and not be penetrated.

    This distance is also shorter than the large leaves,
1000 m (because in that location the tank is 2 cm thicker)

The really extra smart ones
    will do the same for 11 and 1 o'clock
    The distance is
1000 m shorter than the smaller leaf,
    (because the TIGER is 2 cm thicker in that location than it is in front).

Moral: Should one of them in Your cloverleaf be met, then throw him out with a pirouette.

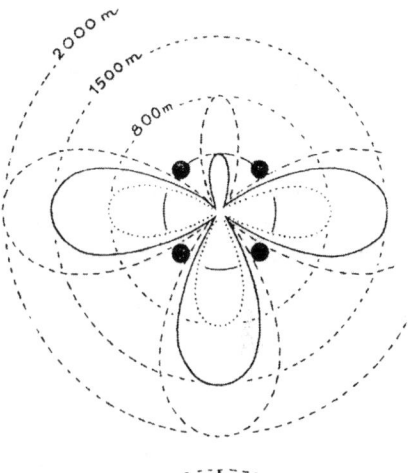

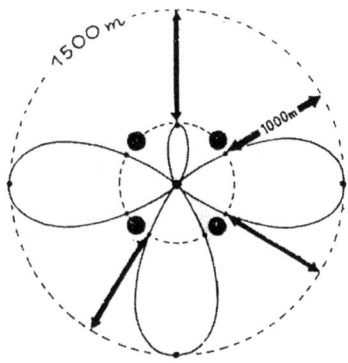

The Reticule Gap

Motto: The average one will shoot a lot, the master shoots with chart and plot.

Moral: The reticule-gap will show You then that You can crack him and when.

The artist compares his model with his work! If the sculpture fits right in between the two tips of the gauge, then he knows that he has the correct measurement.

The tanker compares the enemy to the reticule! If the T-34, seen from the front, fits exactly in between two reticules, then he has the correct measurement to fire at. You will then know:

1. **That You will now penetrate him, and**
2. **which distance that is.**

Through side and rear You can crack all enemy tanks under 2000 m. That is easy to remember.

All of them have a thicker front. In that case You must go closer, or let them get closer, for the T-34 that is 800 m. This distance is different for all tanks. Study the chart on armor location supplied with the manual!

The reticule-gap will tell You when You are close enough to shoot. For the T-34 it is 43, for instance.

4 = reticule-gap, front: The T-34 must be 4 notch wide, so that You may kill him head on. (He has to fit between two reticules). He is then 800 m distant.

3 = reticule-gap, side: The T-34 must be 3 notch wide, so that You can penetrate the side. He is then 2000 m distant.

reticule-gap, rear: is always half of reticule-gap, side, in this case 1½ notch. He is then 2000 m distant.

For really smart ones:
If an enemy tank turns from "side" to his own "mealtime", then his target size will be enlarged by 10% at the most. This error of 10% is included. You must then shoot at the turret center, so that Your shot will hit vertically.

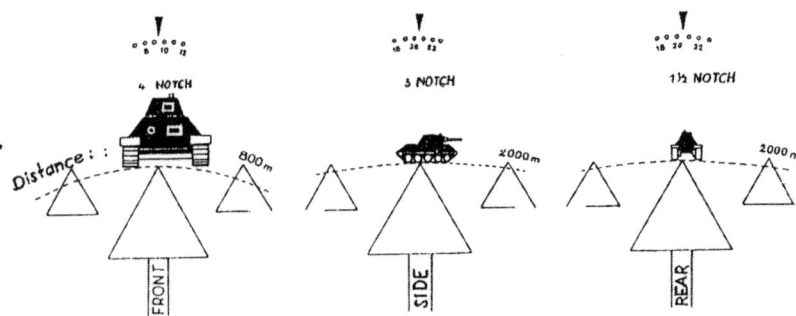

Motto: This reference You must remember like Your bride's picture and number.

Every kid knows the Spitfire and the HE 111.

Every youngster can tell a Ford V-8 from an Opel Kapitaen at 500 m distance. The old foxes recognize a DKW-250 by its sound.

Surely then You must be able to learn the differences between each enemy tank and learn to recognize them! Immediately sit down and study the tank identification chart supplied with this manual.

Memorize the appearance and the following 5 character references:

T - 34	15	8	43
KW - 1	9	4	84
Churchill III	7	15	24
Lee	8	20	13
Sherman	8	8	43

"Wanted"

Reference: Theo XXXIV, a.k.a. T-34 to be shot on sight!

characteristic features

helpful hints:
T-34   15   8   43

You will then master the tank-duel with each of the enemies in Your sleep.

T - 34 type	15 clover-leaf	8 distance	4 reticule-gap, front	3 reticule-gap, side
Your poor enemy	I will be penetrated side and rear closer than 1500 m. Front always 1000 m less, in this case on from 500 m, never at "mealtime"	I penetrate front at 800 m side and rear at 2000 m for all tanks.	4 notch is width of T-34 at 800 m.	3 notch is width of T-34 at 2000 m, rear is always half of reticule-gap, side in this case 1½ notch

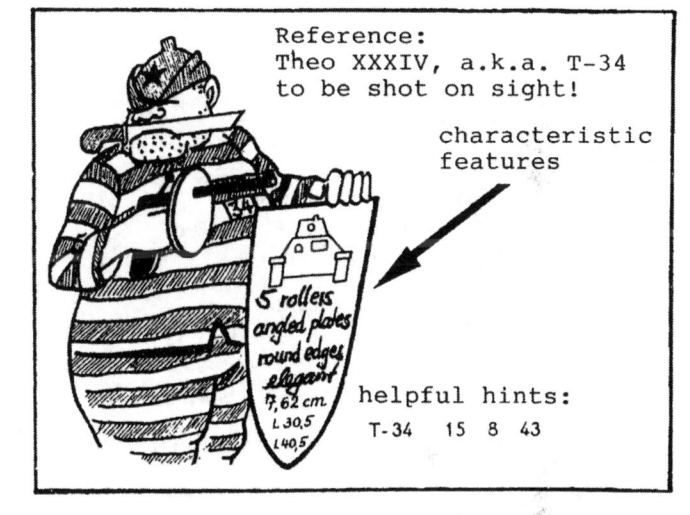

Moral: Often the same number is fatal or full of bliss. Do You stand to lose or gain? Who then, whom will reign.

Anti-Goetz

> **Motto:** He who reaches farther easily can knock the other one in the pan!

You hold the enemy at a distance with Your iron fist and knock him out without him being able to even nibble on You.

You are further away from the other guy than he is from You!

The Anti-Goetz is the space between Your cloverleaf and Your maximum range.

Pay attention!   You can kill the T-34 head on at 800 m. The T-34 can kill You only starting at 500 m.
Anti-Goetz: **Between 500 m and 800 m You can kill him, but he cannot kill You.** You must try to be in that range for battle!

If You stand at "mealtime", You cannot possibly lose this round!
You take more weight and more range into the ring.
You always beat him!
Isn't that a killer thing?

The 5 charts supplied with the manual lay out Your prospects in a tank-duel. They contain cloverleaf, reticule-gap, character reference and Anti-Goetz for Your most dangerous enemies. Look at them for however long it takes to memorize them, so that You will know the entire chapter as soon as their description is on Your lips. Just the same, as You do with the picture in Your shirt-pocket, when thinking of "her".

It's the moral of the story

I can You, but You can't me

Tank Theft

For every shell, that You fire off
    Your father has paid 100 RM in taxes,
    Your mother worked one week in the
    factory.
    the trains had to go 10,000 km.

Think of that each time You want to fire!

Explosive shells onto targets not positively identified or onto targets that can be killed with the machine gun are a crime.

Anti-tank grenades fired at useless range, at tanks already damaged, or poorly aimed only chisel off on steel!

Men of the TIGER! Save!
    Make good use of the thick armor! Go for it!
    It's cheaper to roll over than to shoot the machine gun! Machine gun is cheaper than cannon! Return the cases and packaging material.

The TIGER guzzles down the fuel by the can.
Every Liter has to be hauled from 3000 km away:

Men of the TIGER! Save!
    Be stingy with every Liter!
    Don't let the engine run uselessly!
    Do You know when the next fuel ration will arrive?

All included the TIGER costs 800,000 RM and 300,000 man-hours to produce.
30,000 people have to give a whole weeks pay and 6000 men have to work a whole week, so that You will get one TIGER. They all work for You.
Men of the TIGER!
    Consider what You hold in Your hands! Keep it in good shape!
    They are stealing the tanks! Grab them where the thieves are found!

Supplied with this manual are:

1. tank identification chart Russia

2. armor location chart 8.8 KwK 36

3. Anti - Goetz   T - 34

                 K V 1

                 Churchill III

                 Lee

                 Sherman

Their Engagement
hereby proclaim:
PFC. Barrelbum
and Mrs.
Elvira, born Tiger

Read the following chapters first:

Daily Meals..................page 90

The Cloverleaf...............page 92

The Reticule Gap............page 94

"Wanted".....................page 95

The Belly Button Rule.......page 77

and the Anti-Goetz..........page 96

T34 8 43

REAR

SIDE

2000m
1900
1800
1700
1600
1500
1400
1300
1200
1100
1000

FRONT

9 12 3
   6

"ANTIGÖTZ" I can shoot you, but you can't reach me.

No trespassing into cloverleaf by T34

Read the following chapters first:

Daily Meals.................page 90

The Cloverleaf..............page 92

The Reticule Gap............page 94

"Wanted"....................page 95

The Belly Button Rule.......page 77

and the Anti-Goetz..........page 96

KV I   9 4 84

REAR

SIDE

FRONT

2000 1900 1800 1700 1600 **1500** 1400 1300 1200 1100 1000 900 800 700 600 500 **400** 300 200 100

12 9 3 6

"ANTIGOTZ"
I can shoot you, but you can't reach me!

Trespassing into the cloverleaf by **KV I** is prohibited!

Read the following chapters first:

Daily Meals...................page 90

The Cloverleaf...............page 92

The Reticule Gap............page 94

"Wanted".....................page 95

The Belly Button Rule.......page 77

and the Anti-Goetz..........page 96

Churchill 7 15 24

REAR

FRONT

SIDE

2000 m 1900 1800 1700 1600 **1500** 1400 1300 1200 1100 1000 900 800 **700** 600 500 400 300 200

"ANTIGÖTZ" I can shoot you, but you can't reach me!

Trespassing into the cloverleaf by **Churchill** is prohibited!

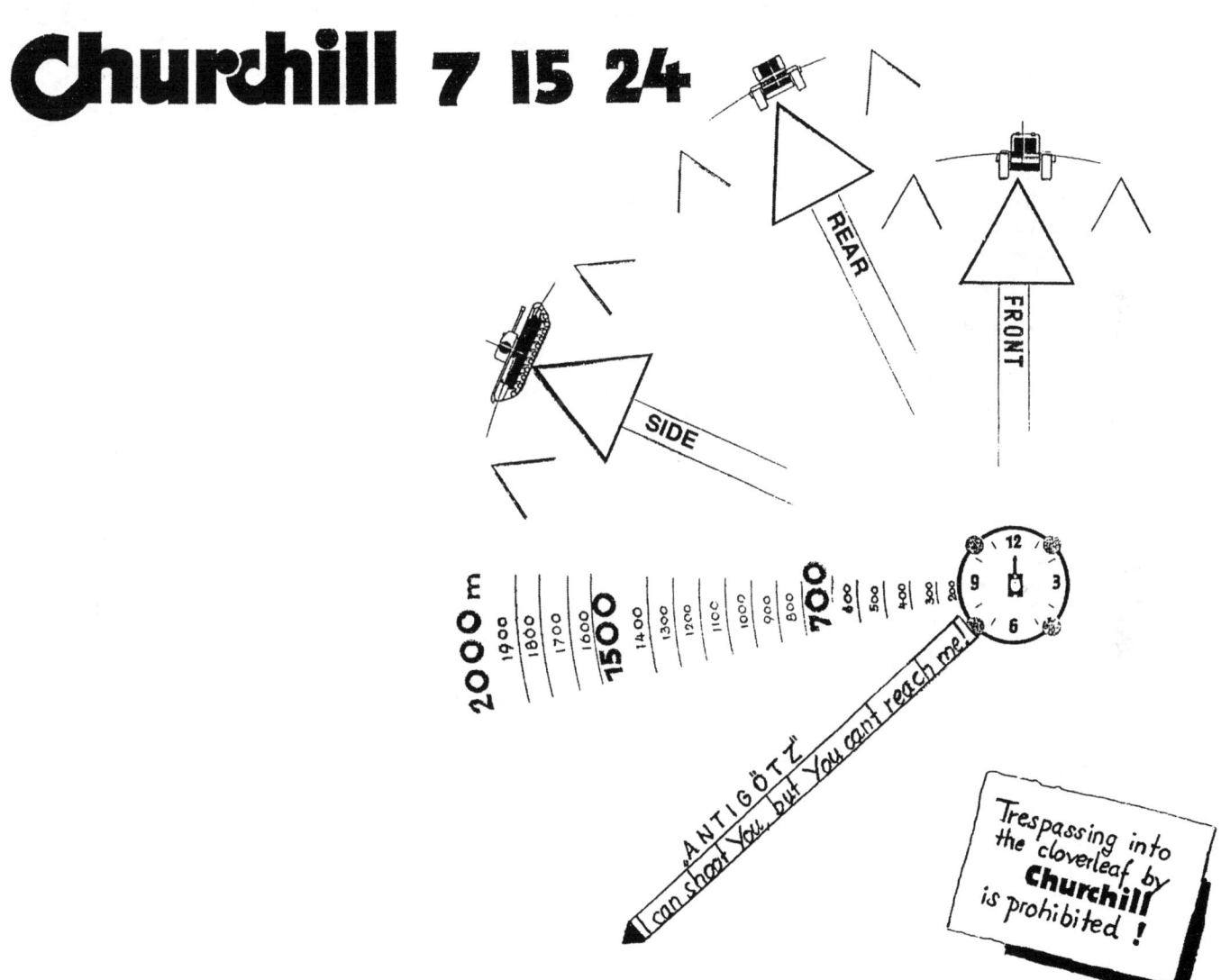

Read the following chapters first:

Daily Meals................page 90

The Cloverleaf............page 92

The Reticule Gap..........page 94

"Wanted"..................page 95

The Belly Button Rule.....page 77

and the Anti-Goetz........page 96

Lee 8 20 15

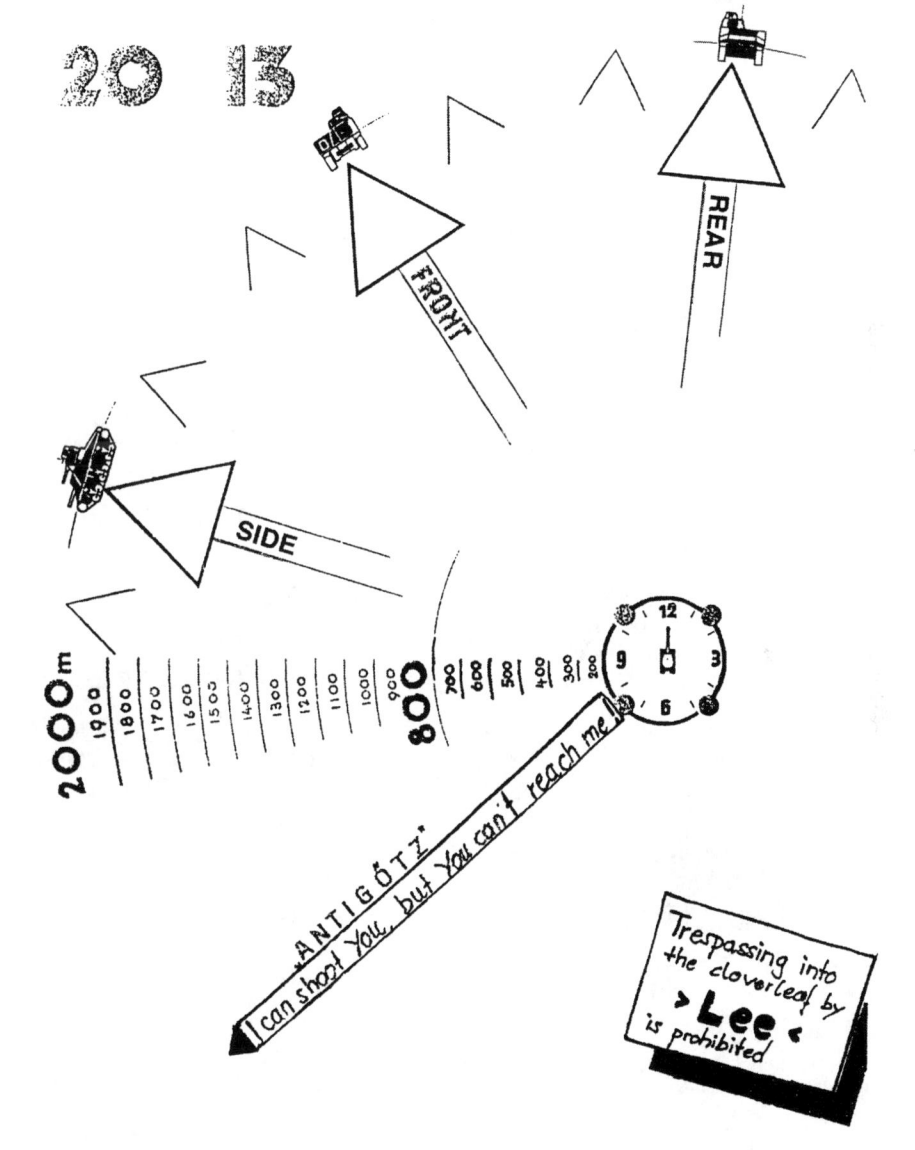

FRONT

REAR

SIDE

2000m 1900 1800 1700 1600 1500 1400 1300 1200 1100 1000 900 800 700 600 500 400 300 200

"ANTIGÖTZ" I can shoot you, but you can't reach me!

Trespassing into the cloverleaf by > Lee < is prohibited

General **Sherman**
6 rollers
round edges
agile
fast
7.5 cm
L/31

Read the following chapters first:

Daily Meals..................page 90

The Cloverleaf..............page 92

The Reticule Gap...........page 94

"Wanted"....................page 95

The Belly Button Rule......page 77

and the Anti-Goetz.........page 96

Sherman 8 8 43

"ANTIGÖTZ"
I can shoot you, but you can't reach me!

Trespassing into the cloverleaf by **Sherman** is prohibited!

114

Supplement to
H.Dv. 469/2a

Armor Intelligence Chart 2

(chart for instruction)

The soviet-russian Armored Weapon

Issued 2/1/1943

(chart for instruction dated 7/1/1942 is to be destroyed)

Important Designs	Can be counted on to appear frequently

Light Armored Vehicles

Armored vehicle **T 60**

Weight: 5.5 t	Weaponry: 1 20mm mechanized cannon

Armor:

Hull and superstructure	Turret
Bow..............20 mm	Shield.....15 mm
Driver's side.....20 mm	Front......15 mm
Side.............15 mm	Side.......15 mm
Rear..........10-13 mm	Rear.......15 mm
Roof.............10 mm	Roof........7 mm
Floor..........7-10 mm	

Weaponry: 1 20mm mechanized cannon
         1 machine gun.
Crew of 2
Dimensions: 4.00 m long, 2.35 m wide
        1.80 m high
All terrain capability:
        ascends 0.55 m,
        crosses 1.40 m,
Ground clearance: 0.30 m
Range: Road 615 km, terrain 315 km
Speed: 44 km/h

OBVIOUS FEATURES:
Uses the same running gear and drive line as the T 40 and differs from it in appearance only in the shape of the upper armor case and the missing propeller drive in the rear.

Use:
Light armored vehicle for battle reconnaissance, usually comprise the (3.) light company of an armored detachment

Evaluation:
Very small and light armored vehicle of most recent production, (used since Nov. 1941), weak armor, minimal combat use. Has mechanical limitations.

Armored vehicle T 70

Weight: 9.2 t
Armor:

Hull and superstructure		Turret	
Bow	45 mm	Shield	60 mm
Driver's front	35 mm	Front	35 mm
Side	16 mm	Side	35 mm
Rear	25 mm	Rear	35 mm
Roof	10 mm	Roof	10 mm
Floor	10 mm		

Weaponry: 1 45mm mechanized cannon L/46, 1 machine gun, 1 machine pistol.
Crew of 2
Dimensions: 4.29 m long, 2.33 m wide, 2 m high
All terrain capability:
  ascends 0.65 m,
  crosses 1.80 m,
  fords 0.90 m
Range: Road 450 km, terrain 300 km
Speed: 45 km/h

OBVIOUS FEATURES:
Outer appearance similar to the T 60, but longer and more imposing stance. Has 5 rollers and a sharply protruding mantlet for cylindrical mount.

Use:
Light armored vehicle for battle reconnaissance, usually comprise the (3.) light company of an armored detachment.

Evaluation:
Light armored vehicle of most recent production. (In use since summer of 1942).

Light Armored Vehicles

Armored vehicle T 27

(Series denomination T 27 A, small armored vehicle, "tankette", armored tractor)

T 27

Weight: 1.7 up to 2.7 t

Armor: 4 up to 10 mm

Weaponry: 1 machine gun

Crew of 2

Dimensions: 2.60 m long, 1.80 m wide, 1.45 m high
All terrain capability: ascends 0.50 m, crosses 1.30 m, fords 0.70 m
Ground clearance: 0.34 m
Range: Road 110 km, terrain 60 km
Speed: 40 km/h

OBVIOUS FEATURES:
Low, boxlike armored vehicle, appears almost as wide as long.

Use:
Originally intended as armored reconnaissance vehicle it has not proven to be feasible as such and is now - often unarmed - used only as an armored artillery tractor.

Evaluation:
Useless as an armored vehicle. Only limited use as a tractor (weak motor).

Amphibious Armored Vehicles

3-Ton Amphibious Armored vehicle

T 37

5-Ton Amphibious Armored vehicle

T 37 and T 38
T 40

O B V I O U S F E A T U R E S :
Low, flat, boxlike construction. Propeller drive in the rear.

O B V I O U S F E A T U R E S :
Low, flat, boxlike construction.
Propeller drive in the rear.

Use:
Light armored vehicle for combat reconnaissance and attack with infantry. Also used commonly when the crossing of waterbodies is not expected.
In terms of overall battle formation amphibious armored vehicles were deployed in large numbers within the Soviet-Russian armored force.

Evaluation:
Weak armor and weaponry, only limited combat use, easy to combat. In light of the few existing bridges in the East, this vehicle's amphibious capability can prove to be of advantage. T 37 and T 38 are obsolete pre-war designs.

T 37 and T 38

Weight: 3.2 t
Armor: turret, Bow, Armor case
   10 mm, otherwise **4** to **6** mm
Weaponry: 1 machine gun
     (T 38 on occasion
       heavier weapons)
Crew of 2
Dimensions: 3.75 m long, 2.00 m wide,
            1.80 m high
Ground clearance; 0.30 m
Range: Road 185 km, terrain 115 km

T 38

T 40

Weight: 5.5 t
Armor: 10 up to **14 mm**, turret,
     bow, armor case otherwise
     6 mm
Weaponry: 1 extra heavy machine gun,
         1 machine gun
Crew of 2
Dimensions: 4.10 m long, 2.35 m wide,
            1.95 m high
Ground clearance: 0.34 m
Range: Road 360 km, terrain 185 km

All terrain capability: ascends 0.60 m,
                       crosses 1.70 m,
                       floats
Speed: on land 45 km/h, in the water
       5 - 10 km/h

T 40

T 26

Armored vehicle

Series denomination: T 26 A
T 26 B
T 26 C

Side view T 26 C

Weight: 9.5 t
Armor:
  Bow and driver's front....16 mm
  Turret....................16 mm
  Side......................16 mm
  Rear......................16 mm
Weaponry: 1 45mm mechanized cannon,
  2-3 machine guns
	Crew of 3
	Dimensions: 4.60 m long, 2.45 m wide,
	  2.65 m high
	All terrain capability: ascends 0.80 m,
	  crosses 2.20 m,
	  fords 0.80 m
	Ground clearance: 0.37 m
	Range: Road 350 km, terrain 175 km
	Speed: 30 km/h

Various turret shapes

Series A 2 turrets   Series B   Series C

T 26

OBVIOUS FEATURES:
Rear of turret on versions B and C cantilevers far out (Turret shape see illustration)
Low type of construction, descending towards the rear.

Use:
Used to be a common light Soviet-Russian armored vehicle for an attack (in conjunction with infantry). After heavy losses in the summer of 1941 it is only rarely seen.

Evaluation:
Weak armor. Good weaponry. Weak engine, slow speed. Obsolete prewar design that is easy to combat. Further production likely ceased.

Flamethrower armored vehicle
T 26 B with flamethrowing device

Turret cutouts

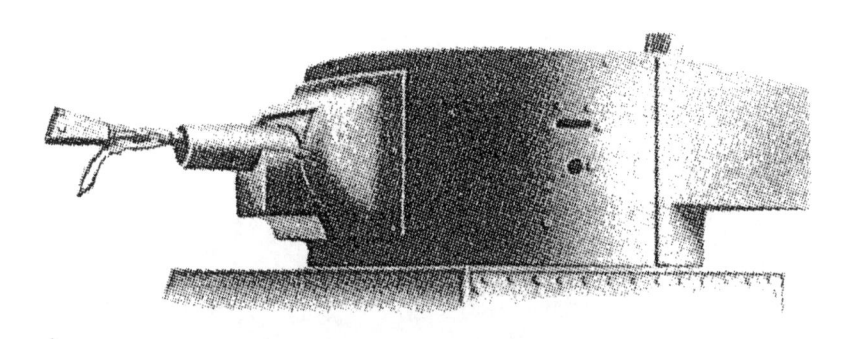

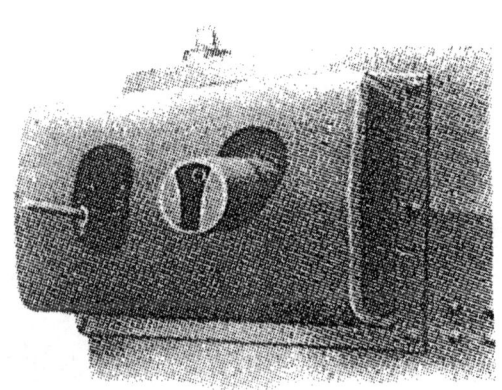

Version 1 (more numerous than version 2)

Version 2

same as light armored vehicle T 26 B, however:
Weaponry: Flamethrowing device
(400 Liters of flame oil),
1 machine gun.
Crew of 3

Obvious features:
Wide box-shaped container on the front of the turret with two openings, one for the machine gun, one for the flame bucket.

Use:
Light armored vehicle for attack with infantry support, especially onto fortified positions. In terms of battle formations intended for use in larger numbers, yet use is comparatively rare.

Evaluation:
Characteristics same as light armored vehicle T 26, limited tactical value.

Armored vehicle

BT

BT 7

"BT" = bystrochodni tank
     = fast armored vehicle
Christie armored vehicle BT 1-7 (various series)

Weight: 12.2 up to 13.7t
(varies by series)

Armor:
  Bow and driver's front......13-22 mm
  Turret........................15 mm
  Side..........................13 mm
  Rear..........................13 mm

Weaponry: 1 45 mm mechanized cannon
  1 machine gun.
  (some have additional machine gun
    in rear of turret)
  (BT 7 sometimes has short 76.2mm
    motorized cannon)

Crew of 3
Dimensions: 5.80 m long, 2.30 m wide,
            2.40 m high
All terrain capability: ascends 0.75 m,
                        crosses 2.10 m,
                        fords 1 m
Ground clearance: 0.36 m
Range: on tracks 430 km, on wheels 570 km
Speed: on tracks 50 km/h,
       on wheels 70 km/h

O B V I O U S   F E A T U R E S :

Large roadwheels (Christie system), flat construction, pointed towards the front (shape of turret changed frequently). The BT 7 armored vehicle equipped with the 7.62 cm motorized cannon is very similar in appearance to the somewhat wider and heavily armored T 34. It is often confused with the T 34. Differences to the T 34 see there.

Use:
Light armored vehicle on wheels (rare) for tactical and operational reconnaissance. On tracks (almost exclusively), used for attack with infantry support. Was used in the Soviet-Russian tank corps in large numbers, together with the T 26 originally about 3/4 of all armored vehicles. Only rarely seen now.

Evaluation:
Fast, light armored vehicle with good pickup, however armor is insufficient. Production likely ceased.

126

Light Armored Vehicle of American Origin

the use of which can be counted on in the Soviet-Russian scenario of war

Armored vehicle

M 3

General Stuart

Weight: 13 t

Armor:

Hull and superstructure	Turret
Bow (cast).........50 mm	Shield (cast)
Driver's front....38 mm	43 mm
Side..............25 mm	Front...55 mm
Rear..............25 mm	Side....32 mm
Roof..............10 mm	Rear....32 mm
Floor.............10 mm	Roof....12 mm

Weaponry: 1 37mm mechanized cannon
Crew of 4
Dimensions: 4.46 m long, 2.46 m wide, 2.65 m high
All terrain capability: ascends 0.65 m, crosses 1.80 m, fords 1.10 m
Ground clearance: 0.42 m
Range: 160 km
Speed: 56 km/h

OBVIOUS FEATURES:
Short, high construction. The front drive wheel is star shaped. Three different turret shapes: a.) round, without commander's cupola elevated, b.) round with commander's cupola elevated, c.) angular, with commander's cupola elevated. Guard for tracks on the side only on version for the tropics.

Use:
Light armored vehicle for tactical and operational reconnaissance.

Evaluation:
Fast, light and agile armored vehicle with good pickup. A lot of wear on the tracks.

Medium Armored Vehicles

Armored vehicle **T 34**

Series denomination.
T 34 A
T 34 B
T 34 B with cast turret

```
Weight: 26.3 t
Armor: T 34 A                                  Armor: T 34 B
    Hull and superstructure    Turret              Hull and superstructure    Turret
    Bow..............45 mm     Shield              Bow..............45 mm     Shield 45+25 mm
    Driver's front....45 mm        45+25 mm        Driver's front 45+15 mm    Front..45+17 mm
    Side..........40-45 mm     Front 45 mm         Side.............45 mm     Side...45+17 mm
    Rear.............40 mm     Side  45 mm         Rear.............45 mm     Rear......45 mm
    Rooif.........18-22 mm     Rear                Roof..........18-22 mm     Roof......16 mm
    Floor............14 mm         40-45 mm        Floor............14 mm
                               Roof  16 mm
```

For the newest version, the driver's front is to be reinforced to a strength of 100 mm.

Armor: T 34 B with cast turret

Hull and superstructure		Turret	
Bow	45 mm	Shield	45+25 mm
Driver's front	45+15 mm	Front	60-70 mm
Side	45 mm	Side	60-70 mm
Rear	45 mm	Rear	60-70 mm
Roof	18-22 mm	Floor	20 mm
Floor	14 mm		

Weaponry: 1 motorized cannon 7.62 cm, 2 machine guns.
Crew of 4
Dimensions: 5.90 m long, 3.00 m wide, 2.45 m high
All terrain capability: ascends 0.90 m, crosses 3.00 m, fords 1.10 m
Ground clearance: 0.38 m
Range: Road 450 km, Terrain 260 km
Speed: 50 km/h

OBVIOUS FEATURES:

Flat construction, angled bow, Christie-running gear (road wheels). Lead vehicles with longer 76.2mm mechanized cannon L/41.5, all other vehicles equipped with shorter mechanized cannon L/30.5.

Features that are different on T 34 as opposed to BT 7, which has a similar appearance:

T 34
Bow-plate: round edges, upper bow-plate including driver's front at shallow angle **(30° incline from horizontal)**
Armor case: angled surfaces
Turret shape: round edges, angled shape
Running gear: 5 road wheels

BT 7
Bow-plate: sharp edges, steep, seperate from driver's front.
Armor case: steep surfaces
Turret shape: steep sides, sharp edges
Running gear: 4 road wheels

In general, the T 34 has a more imposing and smoothed-out (streamlined) appearance.

Use:
Most important armored vehicle for armored attack.

Evaluation:
By far the best and most practical Soviet-Russian armored vehicle. Fast, agile, very powerful weaponry and strong armor. Most difficult to combat of all Soviet-Russian designs, due to its very beneficial construction (incline from horizontal is 30° at bow-plate, 40° - 45° at rear plate, 50° at armor case sides).

Details on attack compare "H.Dv. 469/3"

M 4

Armored vehicle

General Sherman

Weight: circa 31 t

Armor:
Hull and superstructure		Turret	
Bow	55 mm	Shield	244 mm
Driver's front	65 mm	Front	85 mm
Side	26–39 mm	Side	85 mm
Rear	26–60 mm	Rear	85 mm
Roof	13–26 mm	Roof	30 mm
Floor	14–18 mm		

Weaponry: 1 75mm mechanized cannon,
 1 machine gun
 in rotating turret,
 1 anti-aircraft machine
 gun, 1 machine gun in bow
Crew of 5
Dimensions: 5.65 m long,
 2.75 m wide,
 2.75 m high
All terrain capability:
 ascends 0.70 m
 crosses 2.20 m
 fords 1.28 m
Ground clearance: 0.38 m
Range: Road surface 300 km,
 Terrain 160 km
Speed: 36 km/h

Stocky appearance, rounded off design on all sides.

Use:
Most important USA armored vehicle for armored attack and infantry support.

Evaluation:
Most practical USA armored vehicle. Fast, agile, heavy armor and powerful weaponry.

Medium Armored Vehicle of American origin
the use of which can be counted on in the Soviet-Russian scenario of war

Armored Vehicle M 3
"General Lee I"

General Lee

```
Weight: 28 t
Armor:
  Hull and superstructure    Turret
    Bow..............65 mm    Shield..55 mm
    Driver's front....50 mm  Front...88 mm
    Side.............38 mm   Side 50-60 mm
    Rear..........26-38 mm   Rear....50 mm
    Roof.............14 mm   Roof....30 mm
    Floor.........14-18 mm
Weaponry: 1 75mm mechanized cannon L/31
          mounted in bay on one side
          1 37mm mechanized cannon L/56.6
          + 1 machine gun in rotating
          turret,
          1 anti-aircraft machine gun
          mounted in rotating turret-top
          2 machine guns fixed in bow.
```
```
Crew of 7

Dimensions: 5.65 m long, 2.75 m wide,
            3.05 m high

All terrain capability: ascends 0.70 m,
                        crosses 2.20 m,
                        fords 1.28 m

Ground clearance: 0.38 m

Range: Road surface 300 km,
       Terrain 160 km

Speed: 0.38 m
```

"General Lee II" has cast upper part of armor case. Specifications same as "Lee I".

Chassis and superstructure the same as armored vehicle M 3 "General lee I".
Turret without rotating top and cantilevered rear.

"General Grant II" has a cast upper armor case. Specifications same as "Lee I",
but lower, 2.75 m high.

O B V I O U S   F E A T U R E S :;
Elevated construction with bay-like bumpout on the side. Strong weaponry. The front
drive wheel has a star-like shape.

30-ton Armored vehicle M 3 (Canadian)

Another version of the M 3 manufactured in Canada is the armored vehicle "Ram".
It is manufactured of cast armor and features no side bump-out.

Weaponry: "Ram I" 1 40mm mechanized cannon L/52, 2 machine guns
          "Ram II" 1 57mm mechanized cannon L/45, 2 machine guns

Armored vehicle M3
"General Grant I"

General Grant I

Use:
Armored vehicle for armored attack and infantry support.

Evaluation:
Comparatively fast and maneuverable armored vehicle with high firepower.

The armored vehicle M 3's chassis is also used as a self propelled assault gun with 105mm calibre cannon.

Medium Infantry Armored Vehicle of English Origin

the use of which can be counted in the Soviet-Russian scenario of war

26-Ton Armored Pursuit
Mk II vehicle
„Matilda"

Matilda

```
Weight: 26 t                              Weaponry:
Armor:                                      Matilda III C.S. ("close support"=
  Hull and superstructure    Turret         close combat support) 1 76.2 mechanized
  Bow............75-80 mm    Shield 80mm    cannon L/26.5, 1 machine gun,
  Driver's front....80 mm    Front          2 smoke pistols
  Side...........65-70 mm        75-80mm  Crew of 4
  Rear..............55 mm    Side...77mm  Dimensions: 6.00 m long, 2.55 m wide,
  Roof...........16-23 mm    Rear...70mm              2.50 m high
  Floor.............14 mm    Floor..20mm  All terrain capability: ascends 0.60 m,
Weaponry:                                                          crosses 1.80 m,
  Matilda I.II.III                                                 fords 0.80 m
  1 40mm mechanized cannon L/52           Ground clearance: 0.33 m
  1 machine gun, 2 smoke pistols          Range: Road surface 100 km, Terrain 60 km
                                          Speed 23 km/h
```

O B V I O U S   F E A T U R E S :

Angled sides on turret, round edges (cast steel), strong armor on running gear with large cavities (service openings)

Use: Evaluation:
Armored vehicle for infantry support   Heavily armored and especially difficult to
in an attack.   combat from the side. Tactically slow and
                                            cumbersome. Not suitable for operational use.

Details on attack see "H.Dv. 469/3"

Mk III

Armored Pursuit vehicle 16-Ton „Valentine"

Valentine

Weight: 16 t
Armor:

Hull and superstructure	**Turret**
Bow.............60 mm	Shield 26 mm
Driver's front....60 mm	Front 65 mm
Side............60 mm	Side 60 mm
Rear..........16-60 mm	Rear 45-60 mm
Roof..........10-30 mm	Floor 16-20 mm
Floor..........8-20 mm	

Weaponry:
1 40mm mechanized cannon L/52
1 machine gun, 1 smoke pistol

Crew of 3

Dimensions: 5.45 m long, 2.75m wide, 2.25 m high

All terrain capability: ascends 0.70 m, crosses 2.40 m, fords 1.20 m

Ground clearance: 0.42 m

Range: Road surface 150 km, terrain 103 km

Speed: circa 30 km/h

O B V I O U S   F E A T U R E S :
Round turret with vertical sides, running-gear armor (as opposed to Mk II) is missing.

Use:
Same as Mark II

Evaluation:
Lighter than Mk II, but faster and more maneuverable.

Details on attack, see "H. Dv. 469/3"

Medium Armored Vehicle

Armored vehicle

30-Ton **T 28**

T 28

Weight: 28 to 32 t

Armor T 28:
   Bow and driver's front....30 mm
   Turrets....................23 mm
   Side.....................20+7 mm
   Rear......................20 mm

Weaponry: 1 76.2mm mechanized cannon (L/16.5 or L/24), 3 machine guns,(some have additional rear machine gun and anti-aircraft machine gun).

Armor T 28 (reinforced):
   Bow and driver's front 38-52 mm
   Turrets...................53 mm
   Side...................48+7 mm
   Rear......................52 mm

Crew of 6
Dimensions: 7.25 m long, 2.80 m wide,
           2,75 m high
All terrain capability: ascends 0.90 m,
                      crosses 3 m,
                      fords 0.80 m
Ground clearance: 0.43 m
Range: 180 km
Speed: 35 km/h

OBVIOUS FEATURES:
Noticeably large (longer than KV II 52 t), 3 turrets = 1 main turret with mechanized cannon, 2 side turrets with machine guns, armored running gear.

Use:
The T 28 was the main medium prewar-tank for support of light armored vehicles in an attack.

Evaluation:
Obsolete design with numerous mechanical defects. (down time). Slow and difficult to maneuver. Comparatively weak armor for its size, easy to combat. No longer in production. Already mainly used only as an armored artillery tractor.

Heavy Armored Vehicles

44-Ton <u>Armored vehicle</u>
KV I

Series denomination:
KW I A
KW I B
KW I C
KW I S
KW 8

KV I

("KV" = Klim Voroschilov)
Misleading classification:
"52t KV"

```
Weight: 43.5 t                                  Armor: KV I C
Armor KV I A                                       Hull and superstructure        Turret (cast)
   Hull and superstructure        Turret             Bow....75+25-35 mm             Shield 105 mm
   Bow..............75 mm         Shield             Driver's front                 (rolled steel)
   Driver's front...75 mm         60+25 mm              75+25-35 mm                 Front...120 mm
   Side.............75 mm         Front              Side 90 mm, sometimes          Side....120 mm
   Rear.............75 mm         75+25 mm              90 + 40 mm                  Rear....120 mm
   Roof.............35 mm         Side 75mm          Rear.........75 mm             Roof.....40 mm
   Floor............35 mm         Rear 75mm          Roof.........40 mm
                                  Roof 35mm          Floor........35 mm
Armor KV II B (reinforced)                       Weaponry: 1 76.2mm mechanized cannon L/30.2
   Hull and superstructure        Turret                    2 to 3 machine guns
   Bow        75+25-35 mm         Shield          Crew of 5
   Driver's front                 100 mm          Dimensions: 6.80 m long, 3.35 m wide,
      75+25-35 mm                 Front                       2.75 m high
   Side 75, sometimes             75+35 mm        All terrain capability: ascends 0.90 m,
      75+35 mm                    Side                                    crosses 2.80 m,
   Rear.............75 mm         75+30 mm                                fords 1.45 m
   Roof.............35 mm         Rear 75mm       Ground clearance: 0.52 m
   Floor............35 mm         Roof 35mm       Range: Road 335 km, Terrain 200 km
                                                  Speed: 35 km/h
```

The KV I S is a new version of this armored vehicle, featuring reinforced armor, and a motorized cannon 7.62 cm L/41.5.
The KV 8 armored vehicle is equipped with a flamethrower and a motorized cannon 4.5 cm as well as 4 machine guns.

O B V I O U S   F E A T U R E S :

Comparatively small for its size, sleek shape, rear of turret cantilevers far out, (as opposed to KV II 52 t). Additional armor bolted onto turret and chassis as identified by large hex bolts. Armor welded on in places of weaponry impact, only visible from up close.

Use:
Armored vehicle used for fire support in an armored attack. Deployment similar to that of a self-propelled assault gun.

Evaluation:
Armored vehicle with heavy armor and weaponry. Little operational use. May force results in the course of trench-warfare. Especially version B and C is difficult to combat.

Details on attack, see "H. Dv. 469/3"

Heavy Armored Vehicle of American Origin

the use of which can be expected in the Soviet-Russian scenario of war.

Armored vehicle

M 1

60-Ton
"Dreadnought"

Weight: circa 57 t

Armor: .... 75-200 mm

Weaponry:
 1 105mm long mechanized
 cannon in turret
 2 37mm mechanized cannons
 L/56.5
 in driver's front 2 side
 machine guns and 2 machine guns

Crew of 6 - 7
Dimensions: circa 7.00 m long, circa 3.10m
 wide, circa 3.35m high.
All terrain capability: ascends circa 1.50 m
 crosses 3.30 m
 fords 1.20 m
Ground clearance: circa 0.50 m

Range: 220 km

Speed ca. 30 km/h

Stretched out vehicle with armored running gear.

Use:
Heavy armored vehicle for breakthrough and infantry support during an attack.

Evaluation:
Heavy armor and difficult to combat.

143

144

Armored vehicle

("KV" with 15.2 cm motorized howitzer)
(erroneously named "58t", "64t", "70t", etc.)

KV II

KW II

Weight: 52 t

Armor:
    Bow and driver's front....75 mm
    Turret, rolled steel......75 mm
    Side......................75 mm
    Rear......................75 mm
Weaponry: 1 152mm mechanized howitzer, 1-2 machine guns
    Sometimes additional machine gun in rear.

Crew of 6 - 7
Dimensions: 6.80 m long, (incl. weapon 7.20 m), 3.35 m wide, 3.30 m high.
All terrain capability: ascends 0.90 m, crosses 2.80 m, fords 1.45 m
(can only pass over bridges, of high weight rating)
Ground clearance: 0.52 m
Range: Road 280 km, terrain 170 km
Speed: 30 km/h

O B V I O U S   F E A T U R E S :

Steep, cube-shaped turret case, high overall height, heavy weaponry (howitzer!) Tank chassis, hull and running gear are the same as KV 1 44 t.

Use:
Heavy armored vehicle used for fire support in an armored attack. Deployment similar to that of a self-propelled assault gun.
Now only rarely seen.

Evaluation:
Heavy armor. Very substantial firepower, however limited maneuverability. Useful in trench warfare. Even according to soviet-russian evaluation the overburdened chassis was not satisfactory.

Heavy Armored Vehicles

45 - ton armored vehicle T 35

(TP, BS, S II, M II, T 32 older model, AV T 35 A-B-C different versions)

T 35 A

Weight: 45 t
Armor:
    Bow and driver's front 30 mm
    Turrets..............20-25 mm
    Side................23+11 mm
    Rear................22-27 mm
Weaponry: 1 76.2mm mechanized cannon L/16.5 or L/24, 2 45mm mechanized cannon, 6-7 machine guns.

Crew of 5

Dimensions: 9.60m long, 3.20m wide, 3.50 high
All terrain capability: ascends 1.30 m, crosses 4.75 m, fords 1.25 m

Grund clearance: 0.58 m

Range: Road 150 km

Speed: 30 km/h

T 35 C

Weight: 45 t	**Crew of 6**
Armor: up to 60 mm	**Dimensions:** 9.60 m long, 3.20 m wide, 3.50 m high
Weaponry: 1 76.2mm mechanized cannon 1 45mm mechanized cannon, 2 - 3 machine guns	**All terrain capability:** ascends 1.30 m, crosses 4.75 m, fords 1.25 m
	Ground clearance: 0.58 m
	Range: Road 150 km
	Speed 30 km/h

O b v i o u s   f e a t u r e s :

Production varies widely.
Number of turret varies by series.
T 35 A: 5 turrets, (1 main, 4 side turrets).
T 35 C: 2 turrets, some have armored running gear.

Use:
Heavy armored vehicle for armored attack with infantry suppory.

Evaluation:

Obsolete design, big, cumbersome, limited maneuverability. Especially the T 35 A has little combat use in spite of its number of turrets. Both versions are no longer made and will only rarely be seen.

Continued: Important Designs
Can be counted on to appear frequently

Heavy Infantry-Armored Vehicle of English Origin
the use of which can be expected in the Soviet-Russian scenario of war

Armored Pursuit Vehicle

40-Ton Mk IV "Churchill I / II"

Churchill I / II

Weight: approximately 38 t
Armor:
  Hull and superstructure
Bow..............38-75mm
Driver's front
    88+14 - some 88+88mm
Side 14+38 upto 14+64mm
Rear............28-50mm
Floor.............16mm
Top............16-20mm
Weaponry: Churchill I
  in turret: 1 40mm mechanized cannon
        L752, 1 machine gun,
        1 anti-aircraft machine
        gun, 1 smoke pistol
  in driver's front:1 76.2mm mechanized
        cannon L/26.5,
        2 machine pistols

  Turret
Shield..100mm

Front...100mm
Side....100mm
Rear....100mm
Roof..40-50mm

Weaponry: Churchill II
  In turret: same as Churchill I
  in driver's front: 1 machine gun,
    1 flamethrower, fixed, 2 machine pistols
Crew of 5
Dimensions: Churchill I 7.10m long
            Churchill II, 7.60m long
            3.25 m wide, 2.65 m high.
All terrain capability: ascends 1.13m,
                    crosses 2.80m,
                    fords up to 2.40m
Ground clearance: 0.51m
Driving Range: Road 260 km, Terrain 80 km
Speed: 26 km/h

OBVIOUS FEATURES :

Flat, stretched construction. Unprotected running gear. Rounded off shape of turret

Use:
Heavy armored vehicle used for infantry support during an attack.

Evaluation:
Heavy armor and difficult to combat. Slow tactical effect and cumbersome. Not suitable for operational use.

Heavy Infantry-Armored Vehicle of English Origin

Armored Pursuit Vehicle

40 t

Mk IV
"Churchill III"

Churchill III

Weight: approximately 38t
Armor:
  Hull and superstructure   **Turret**
    Bow..........38-75mm   Shield....55 mm
    Driver's front
    88+14 - some 88+88mm   Front.....88 mm
    Side   Side...75-88 mm
    14+38 up to 14+64mm   Rear......75 mm
    Rear.........28-50mm   Roof......20 mm
    Roof.........16-20mm
    Floor...........16mm
Weaponry: Churchill III
  in turret: 1 57mm mechanized cannon L/45,
             1 machine gun, 1 anti-aircraft-
             machine gun, 1 smoke pistol.
  in driver's front: 1 machine gun,
                     2 machine pistols

Crew of 5

Dimensions: 7.10m long, 3.25m wide, 2.65m high

All terrain capability: ascends 1.13m,
                        crosses 2.80m,
                        fords 2.40m.

Ground clearance: 0.51 m

Range: Road 260 km, Terrain 80 km

Speed: 26 km/h

OBVIOUS FEATURES :
Flat, stretched out design. Unprotected running gear. Angular shape of turret.

Use:
Heavy armored vehicle used for infantry support during an attack.

Evaluation:
Heavy armor and difficult to combat. Slow tactical effect and cumbersome. Not suitable for operational use.

Older Designs

only rarely expected to be seen

Armored Reconnaissance Vehicles

Arm. Rec. Veh. ,,**BA**" (Bronieford)

Arm. Rec. Veh. ,,**BA**" (Ford)

,,BA" Bronieford

,,BA" Ford

"BA" Bronieford	"BA" Ford
Vehicle weight: 1.7 to 2.1t	**Vehicle weight:** 5t
Armor: 5 to 6 mm	**Armor:** 7 to 13 mm
Weaponry: 1 machine gun	**Weaponry:** 1 motorized cannon 4,5 cm, 2 machine guns
Crew of 3 Dimensions: 4.20 m long, 1.70 m wide 2.10 m high Ground clearance: 0.26 m	Crew of 4 Dimensions: 4.70 m long, 2.10 m wide 2.40 m high Ground clearance: 0.22 m

All terrain capability: limited
Driving range: 320 km (road surface)
Vehicle speed: 50 km/h

OBVIOUS FEATURES :
"BA" Bronieford: Two - axle running gear, short edged shape.
"BA" Ford: Three - axle running gear, edged shape. (Turret on new edition of conical shape and rounded off.)

Use:
Light or medium armored reconnaissance vehicle, respectively. Used for tactical and operational reconnaissance by mechanized units.

Evaluation:
"BA" Bronieford: Obsolete version, inferior armor and weaponry.
"BA" Ford: Improved version. Reconnaissance vehicle with combat capability, does however exhibit numerous mechanical shortcomings.

154

Enclosure
H.Dv. 469/3b

F O R   O F F I C I A L   U S E   O N L Y !

Do not let fall in enemy hands!

(Defence against armored vehicles

difficult to combat)

Armor penetration chart

8.8 cm "KwK" 36

Issued: 2 / 15 / 43

Fundamentals on procedure of shooting against armored vehicles difficult to combat

1. Maintain **Coldbloodedness**: Strive to obtain suitable distance to detect "weak spots" and facilitate effective destruction!
2. Combat enemy tanks from **hidden position** and **unexpected direction**! In open terrain, attack enemy tanks "over the corner" of own tank (highest possible protection by the given armor).
3. In spite of careful **aim for each individual shot**, maintain **high firing frequency!**
4. Alert **inspection** of **projectile impact**! Not every hit is always immediately destructive.
5. **Strive to obtain favorable angle of impact!** Highest impact is obtained when front or side are fully visible, the least when traveling at an angle (45°). For round or curved turrets, always hold at center of turret.
6. Select the correct **grade of ammunition**! Consider the data in this chart.
   Solid core shells, use only up to a distance of 2000m and then only if
   regular armor-piercing shells or **HD shells** turn out to have no effect.
   Explosive shells - ignitor position "without delay" - can cause disabling
   favorable circumstances, (setting on fire),
   destructive effect for hitting the front of the turret just above the turret ring or under the cantilevered rear of the turret by means of lifting or diverting installed position of turret on armored vehicles **T 34 A and B**, also for hitting the front of the turret on the **MK II** underneath the cannon.
   These hits on the turret however are rare.
7. In this armor penetration chart the following symbols mean:

Ammunition:	Effect:
At = 88mm armor-piercing shell Kwk 36	■ = Destructive effect
HD = 88mm Heavy Duty shell KwK 36	▨ = Disabling or destructive effect
SC = 88mm #40 solid core sell KwK36	
EG = 88mm inserted explosive cartridge	☐ = No effect

The weak spots of armored vehicles which can be fired at successfully are marked with **Indicating lines** connected to the **abbreviations** of the proper grade of ammunition.

Figures in meters next to the abbreviations for grade of ammunition indicate the **upper limit of the distance** at which a successful penetration of the armor can be counted on. For **HD-grenades** no distances were given, as these projectiles can penetrate all outlined black areas up to a distance of 2000 m. The **combat distance** however, will in most cases be far shorter, when taking into account the size of the target and the surrounding circumstances, (enemy impact, vision, etc.). For details on the effect of different grades of ammunition, study the text part of this manual, (H.Dv. 469/3b).

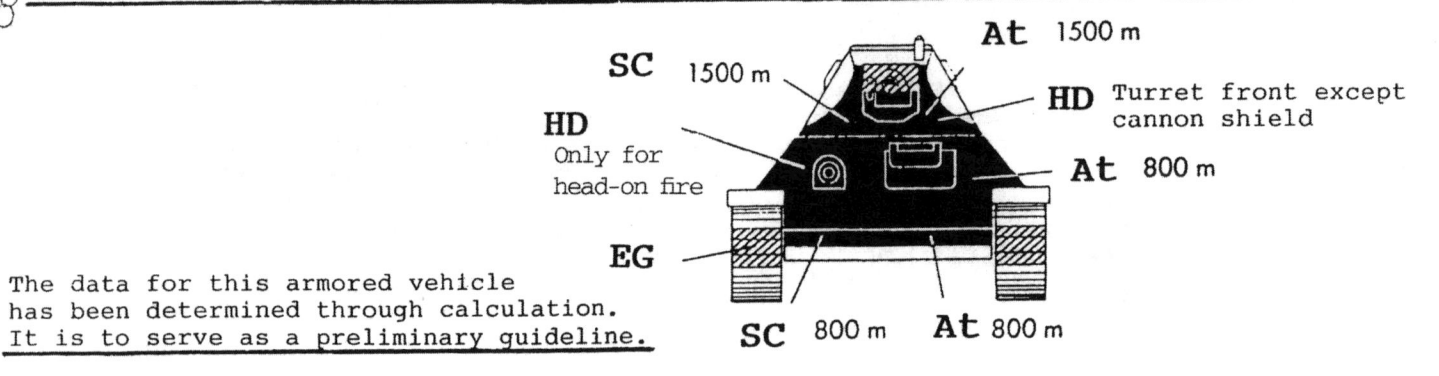

158

SC 1500 m
At 1500 m
HD Turret front except cannon shield
HD Only for head-on fire
At 800 m
EG
SC 800 m
At 800 m

The data for this armored vehicle has been determined through calculation. It is to serve as a preliminary guideline.

Medium armored vehicle

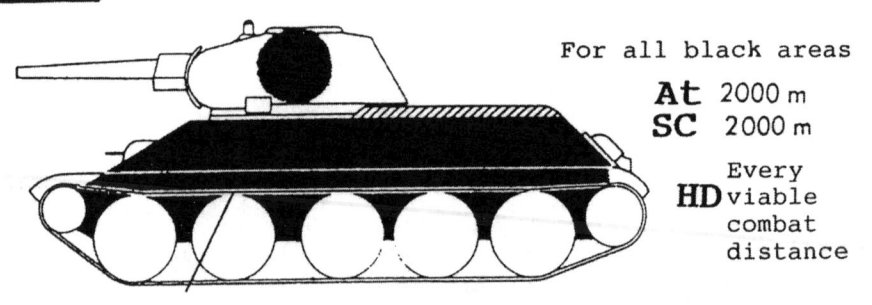

For all black areas
At 2000 m
SC 2000 m
HD Every viable combat distance

EG For firing at track and running gear

T 34 A

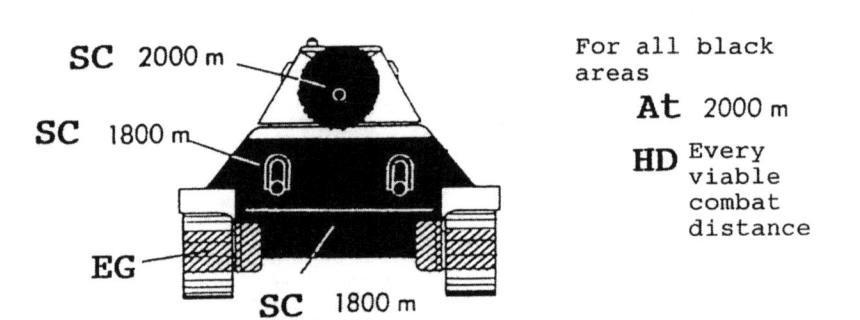

SC 2000 m
SC 1800 m
EG
SC 1800 m

For all black areas
At 2000 m
HD Every viable combat distance

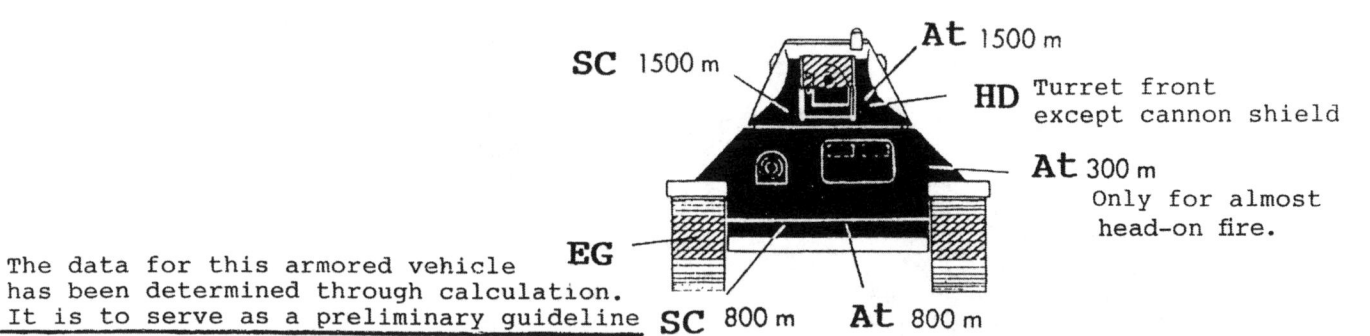

Medium armored vehicle

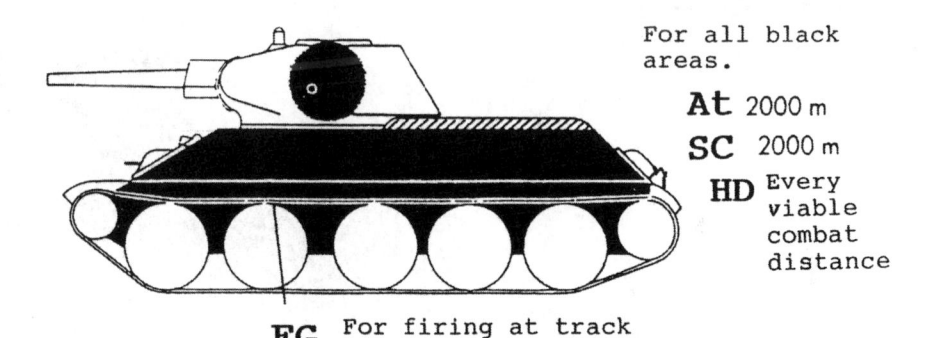

T 34 B

(reinforced)

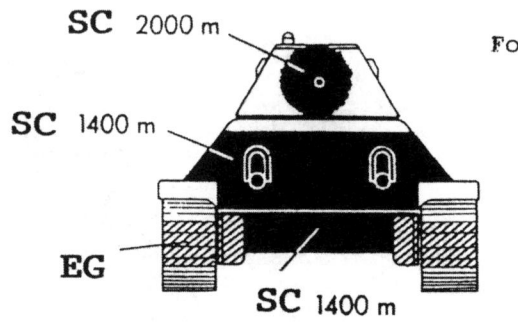

SC 1500 m
At 1500 m
HD Turret front except cannon shield
At 300 m — Only for almost head-on fire.
EG
SC 800 m
At 800 m

The data for this armored vehicle has been determined through calculation. It is to serve as a preliminary guideline

For all black areas.
At 2000 m
SC 2000 m
HD Every viable combat distance

EG For firing at track and running gear.

SC 2000 m
SC 1400 m
EG
SC 1400 m

For all black areas
At 2000 m
HD Every viable combat distance

159

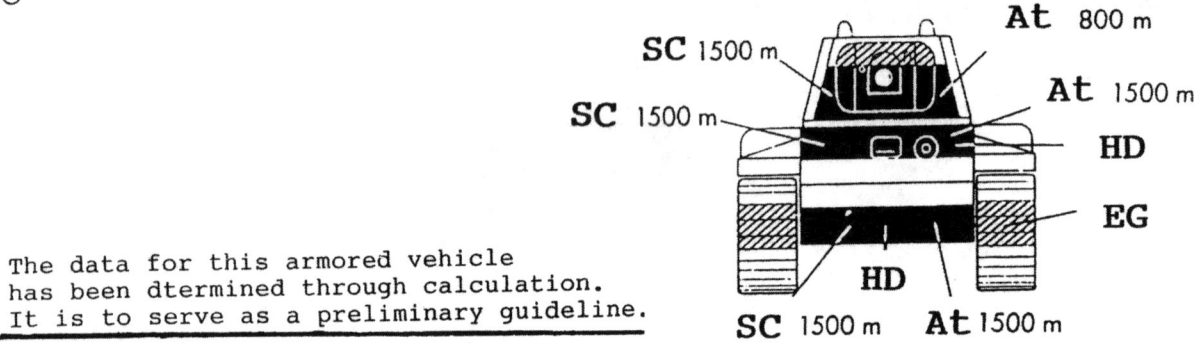

160

The data for this armored vehicle has been dtermined through calculation. It is to serve as a preliminary guideline.

Heavy armored vehicle

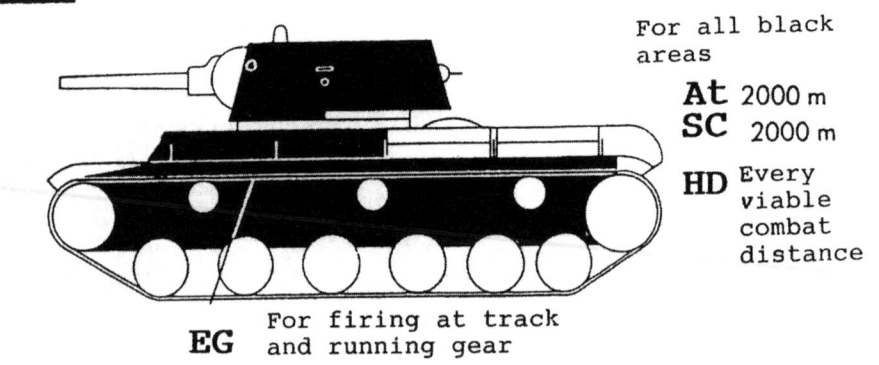

KV I A

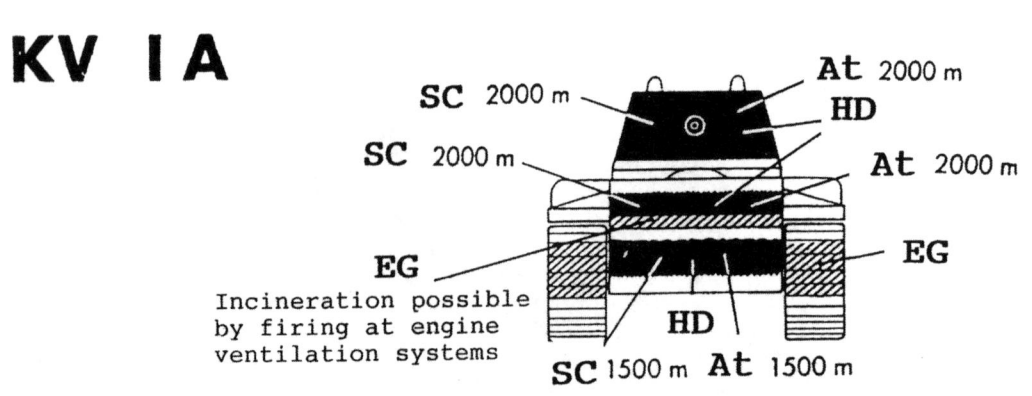

Front view:
- SC 1500 m
- SC 1500 m
- At 800 m
- At 1500 m
- HD
- EG
- HD
- SC 1500 m
- At 1500 m

Side view:
- For all black areas
  - At 2000 m
  - SC 2000 m
- HD Every viable combat distance
- EG For firing at track and running gear

Rear view:
- SC 2000 m
- SC 2000 m
- At 2000 m
- HD
- At 2000 m
- EG Incineration possible by firing at engine ventilation systems
- EG
- HD
- SC 1500 m
- At 1500 m

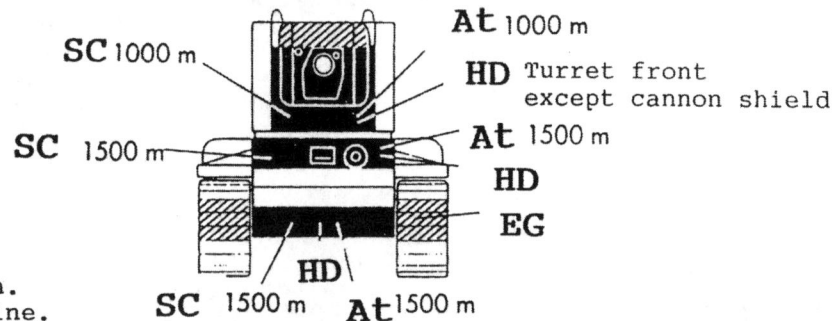

161

SC 1000 m — At 1000 m
HD Turret front except cannon shield
SC 1500 m — At 1500 m
HD
EG
HD
SC 1500 m   At 1500 m

The data for this armored vehicle has been determined through calculation. It is to serve as a preliminary guideline.

Heavy armored vehicle

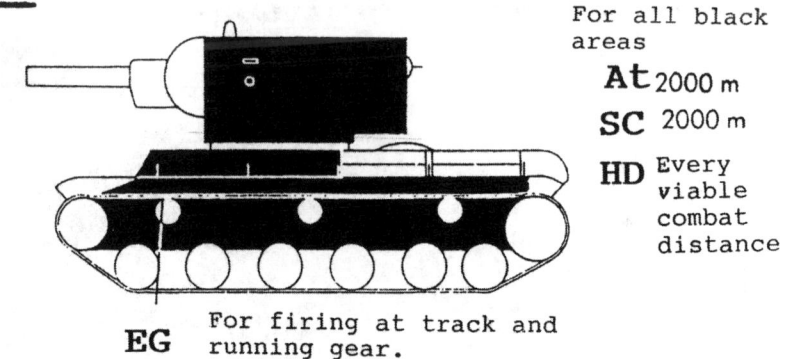

For all black areas
At 2000 m
SC 2000 m
HD Every viable combat distance

EG For firing at track and running gear.

KV II

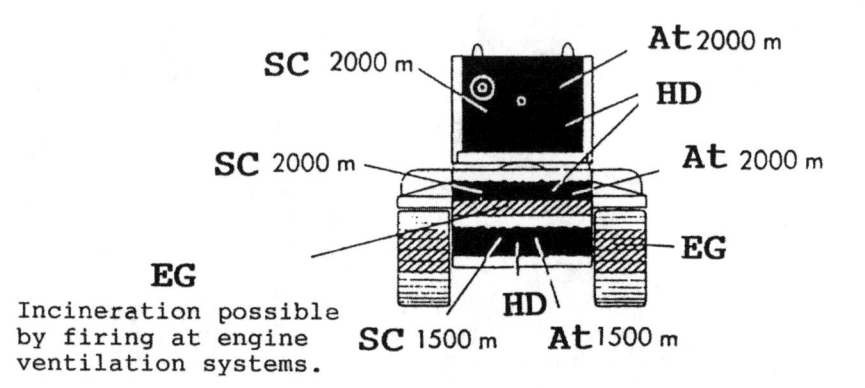

At 2000 m
SC 2000 m
HD
SC 2000 m — At 2000 m
EG
HD
SC 1500 m   At 1500 m

EG Incineration possible by firing at engine ventilation systems.

The data for this armored vehicle
has been determined through calculation.
It is to serve as a preliminary guideline.

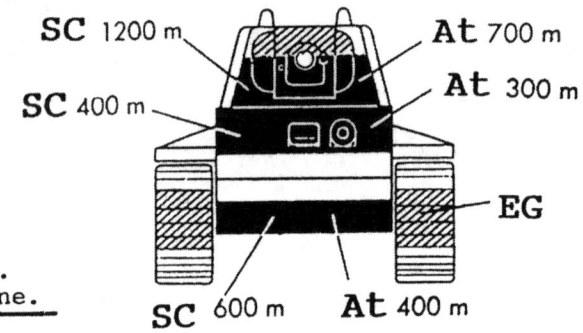

Heavy armored vehicle

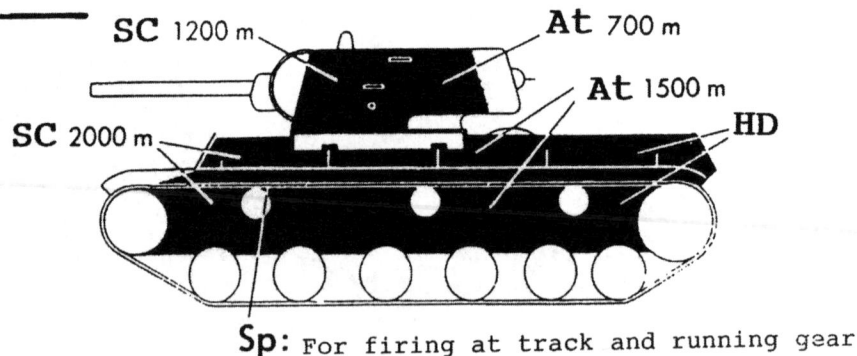

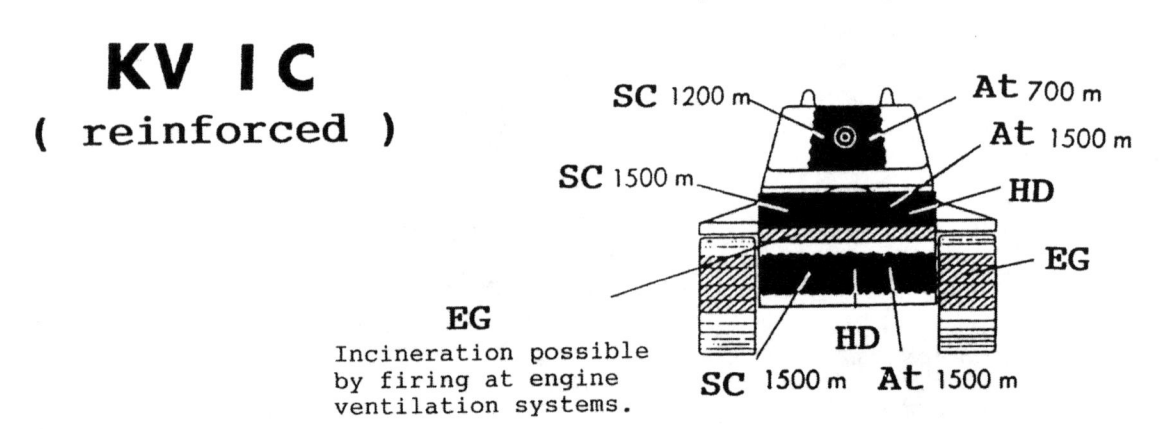

KV I C
(reinforced)

Sp: For firing at track and running gear

EG
Incineration possible by firing at engine ventilation systems.

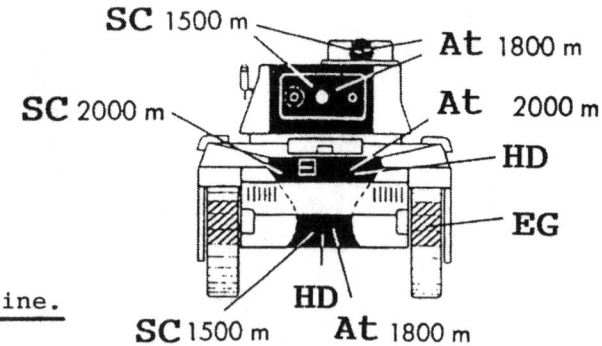

── 163

The data for this armored vehicle has been determined through calculation.
It is to serve as a preliminary guideline.

Medium armored vehicle

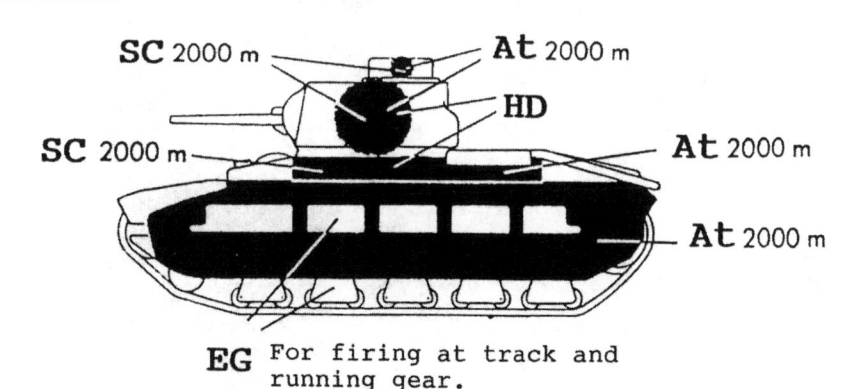

Mk II

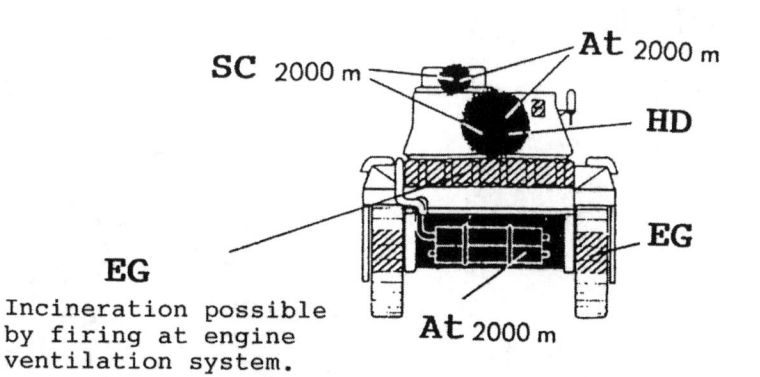

- SC 1500 m
- At 1800 m
- SC 2000 m
- At 2000 m
- HD
- EG
- HD
- SC 1500 m
- At 1800 m

- SC 2000 m
- At 2000 m
- HD
- SC 2000 m
- At 2000 m
- At 2000 m

EG For firing at track and running gear.

- At 2000 m
- SC 2000 m
- HD
- EG
- At 2000 m

EG Incineration possible by firing at engine ventilation system.

The data for this armored vehicle has been determined through calculation. It is to serve as a preliminary guideline.

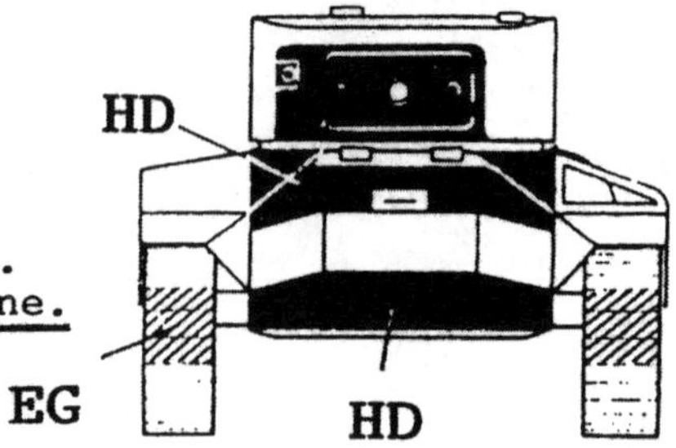

HD
EG
HD

For all black areas

At 2000 m
SC 2000 m

Medium armored vehicle

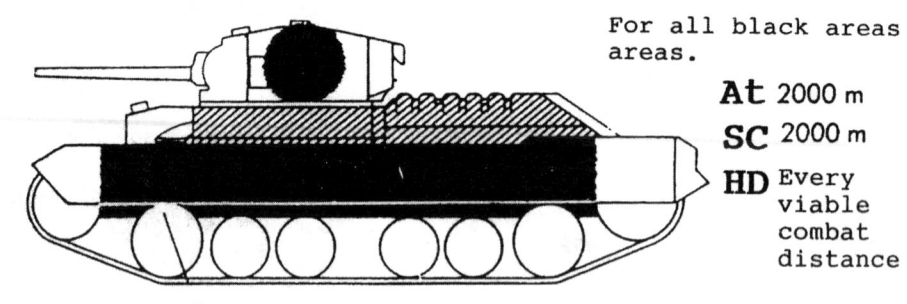

For all black areas areas.

At 2000 m
SC 2000 m
HD Every viable combat distance

EG For firing at track and running gear.

Mk III (Valentine)

Incineration possible by firing at engine ventilation system.
EG

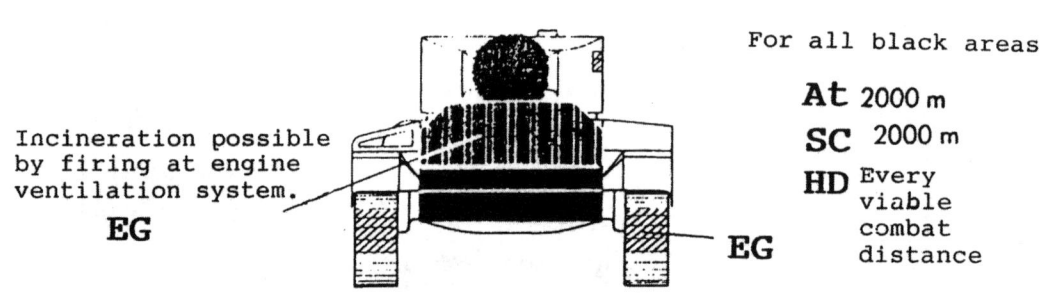

EG

For all black areas

At 2000 m
SC 2000 m
HD Every viable combat distance